Fast Start Integral Calculus

Synthesis Lectures on Mathematics and Statistics

Editor
Steven G. Kranz, *Washington University, St. Louis*

Numerical Integration of Space Fractional Partial Differential Equations: Vol 2 –
Applicatons from Classical Integer PDEs
Younes Salehi and William E. Schiesser
2017

Numerical Integration of Space Fractional Partial Differential Equations: Vol 1 –
Introduction to Algorithms and Computer Coding in R
Younes Salehi and William E. Schiesser
2017

Aspects of Differential Geometry III
Esteban Calviño-Louzao, Eduardo García-Río, Peter Gilkey, JeongHyeong Park, and Ramón
Vázquez-Lorenzo
2017

The Fundamentals of Analysis for Talented Freshmen
Peter M. Luthy, Guido L. Weiss, and Steven S. Xiao
2016

Aspects of Differential Geometry II
Peter Gilkey, JeongHyeong Park, Ramón Vázquez-Lorenzo
2015

Aspects of Differential Geometry I
Peter Gilkey, JeongHyeong Park, Ramón Vázquez-Lorenzo
2015

An Easy Path to Convex Analysis and Applications
Boris S. Mordukhovich and Nguyen Mau Nam
2013

Applications of Affine and Weyl Geometry
Eduardo García-Río, Peter Gilkey, Stana Nikčević, and Ramón Vázquez-Lorenzo
2013

Essentials of Applied Mathematics for Engineers and Scientists, Second Edition
Robert G. Watts
2012

Chaotic Maps: Dynamics, Fractals, and Rapid Fluctuations
Goong Chen and Yu Huang
2011

Matrices in Engineering Problems
Marvin J. Tobias
2011

Fast Start Integral Calculus
Daniel Ashlock

ISBN: 978-3-031-01293-8   paperback
ISBN: 978-3-031-02421-4   ebook
ISBN: 978-3-031-00267-0   hardcover

DOI 10.1007/978-3-031-02421-4

A Publication in the Springer series
SYNTHESIS LECTURES ON MATHEMATICS AND STATISTICS

Lecture #29
Series Editor: Steven G. Kranz, *Washington University, St. Louis*
Series ISSN
Print 1938-1743   Electronic 1938-1751

Fast Start Integral Calculus

Daniel Ashlock
University of Guelph

SYNTHESIS LECTURES ON MATHEMATICS AND STATISTICS #29

ABSTRACT

This book introduces integrals, the fundamental theorem of calculus, initial value problems, and Riemann sums. It introduces properties of polynomials, including roots and multiplicity, and uses them as a framework for introducing additional calculus concepts including Newton's method, L'Hôpital's Rule, and Rolle's theorem. Both the differential and integral calculus of parametric, polar, and vector functions are introduced. The book concludes with a survey of methods of integration, including u-substitution, integration by parts, special trigonometric integrals, trigonometric substitution, and partial fractions.

KEYWORDS

integral calculus, polynomials, parametric functions, polar functions, vector functions, methods of integration

Contents

Preface

This text covers single-variable integral calculus, presuming familiarity with differential calculus from the precursor text *Fast Start Differential Calculus*. The text was developed for a course that arose from a perennial complaint by the physics department at the University of Guelph that the introductory calculus courses covered topics roughly a year after they were needed. In an attempt to address this concern, a multi-disciplinary team created a two-semester integrated calculus and physics course. This book covers the integral calculus topics from that course as well a material on the behavior of polynomial function. The philosophy of the course was that the calculus will be delivered before it is needed, often just in time, and that the physics will serve as a substantial collection of motivating examples that will anchor the student's understanding of the mathematics.

The course has run three times before this text was started, and it was used in draft form for the fourth offering of the course, and then for two additional years. There is a good deal of classroom experience and testing behind this text. There is also enough information to confirm our hypothesis that the course would help students. The combined drop and flunk rate for this course is consistently under 3%, where 20% is more typical for first-year university calculus. Co-instruction of calculus and physics works. It is important to note that we did not achieve these results by watering down the math. The topics covered, in two semesters, are about half again as many as are covered by a standard first-year calculus course. That's the big surprise: covering more topics faster increased the average grade and reduced the failure rate. Using physics as a knowledge anchor worked even better than we had hoped.

This text, and its two companion volumes, *Fast Start Differential Calculus*, and *Fast Start Advanced Calculus* make a number of innovations that have caused mathematical colleagues to raise objections. In mathematics it is traditional, even dogmatic, that math be taught in an order in which no thing is presented until the concepts on which it rests are already in hand. This is correct, useful dogma for mathematics students. It also leads to teaching difficult proofs to students who are still hungover from beginning-of-semester parties. This text neither emphasizes nor neglects theory, but it does move theory away from the beginning of the course in acknowledgment of the fact that this material is philosophically difficult and intellectually challenging. The course also presents a broad integrated picture as soon as possible. The text also emphasizes cleverness and computational efficiency. Remember that "mathematics is the art of avoiding calculation."

It is important to state what was sacrificed to make this course and this text work the way they do. This is not a good text for math majors, unless they get the theoretical parts of calculus later in a real analysis course. The text is relatively informal, almost entirely example driven, and application motivated. The author is a math professor with a CalTech Ph.D. and three decades of experience teaching math at all levels from 7th grade (as a volunteer) to graduate education including having supervised a dozen successful doctoral students. The author's calculus credits include calculus for math and engineering, calculus for biology, calculus for business, and multivariate and vector calculus.

Daniel Ashlock
August 2019

Acknowledgments

This text was written for a course developed by a team including my co-developers Joanne M. O'Meara of the Department of Physics and Lori Jones and Dan Thomas of the Department of Chemistry, University of Guelph. Andrew McEachern, Cameron McGuinness, Jeremy Gilbert, and Amanda Saunders have served as head TAs and instructors for the course over the last six years and had a substantial impact on the development of both the course and this text. Martin Williams, of the Department of Physics at Guelph, has been an able partner on the physics side delivering the course and helping get the integration of the calculus and physics correct. I also owe six years of students thanks for serving as the test bed for the material. Many thanks to all these people for making it possible to decide what went into the text and what didn't. I also owe a great debt to Wendy Ashlock and Cameron McGuinness at Ashlock and McGuinness Consulting for removing a large number of errors and making numerous suggestions to enhance the clarity of the text.

Daniel Ashlock
August 2019

CHAPTER 1

Integration, Area, and Initial Value Problems

The two main concepts in calculus are derivatives and integrals. We assume you have already studied derivatives. This chapter introduces the concept of **integrals**. Derivatives and integrals were developed independently, and later it was discovered that they are closely related. We will start with this relationship and define integrals as a sort of backward derivative. Then, we will introduce some of the primary uses of integration – finding areas under curves and solving boundary value problems. Finally, we will show how integrals were originally defined. This chapter does not cover how to integrate most functions. That we've saved for later in Chapter 4.

1.1   ANTI-DERIVATIVES

The anti-derivative of a function is another function whose derivative is the function you started with. More precisely:

Definition 1.1 *If $f(x)$ and $F(x)$ are functions so that*

$$F'(x) = f(x),$$

we say that $F(x)$ is an **anti-derivative** *of $f(x)$.*

The terminology **anti-derivative** is fairly modern. The original terminology is to call an anti-derivative of $f(x)$ an **integral** of $f(x)$. Integrals also have a special notation.

Knowledge Box 1.1

> **Integral notation**
>
> *If $F(x)$ is an anti-derivative of $f(x)$, we write*
>
> $$F(x) = \int f(x) \cdot dx$$
>
> *and call $F(x)$ an integral of $f(x)$.*

Example 1.1   Find an anti-derivative of $f(x) = 2x$.

Solution:

We know that the derivative of $F(x) = x^2$ is $f(x) = 2x$. So, one possible answer is:

$$F(x) = x^2$$

◊

One of the issues that arises when we work with anti-derivatives is that they are not unique.

The derivative of

$$G(x) = x^2 + 5$$

is *also* 2x. This means that anti-derivatives (integrals) are known only up to some constant value. We will develop techniques for dealing with this in Section 1.3, but for now we will simply use **unknown constants** or **constants of integration**.

Example 1.2   Use integral notation, with an unknown constant C, to represent all the anti-derivatives of $2x$.

Solution:

$$\int 2x \cdot dx = x^2 + C$$

◊

We also need to explain dx. The job of dx is to tell us which symbol is the variable.

Definition 1.2 *The symbol dx is spoken "the differential of x" and is used to designate the active variable in an integral.*

We have seen dx and other differentials before. The symbol $\dfrac{dy}{dx}$ was an alternate way of saying y', for example. There is another use of differential symbols – as an operator.

Knowledge Box 1.2

The differential operator

Another way to say "take the derivative" is with the symbol $\dfrac{d}{dx}$. *This symbol denotes* **the derivative with respect to** x. *So:*

$$f'(x) = \frac{d}{dx}f(x)$$

We will need this symbol, the differential operator, in the next section.

In order to compute $\displaystyle\int 2x \cdot dx$, we used the fact that we knew that $\dfrac{d}{dx}x^2 = 2x$. In fact, each derivative rule is also an anti-derivative rule, just used backward. Let's start by giving the reverse of the power rule for derivatives.

Knowledge Box 1.3

The power rule for integration

$$\int x^n \cdot dx = \frac{1}{n+1}x^{n+1} + C$$

The rules for constant multiples and sums for derivatives also apply for integrals.

Knowledge Box 1.4

Integrals of constant multiples and sums

$$\int a \cdot f(x) \cdot dx = a \cdot \int f(x) \cdot dx$$

$$\int (f(x) + g(x)) \cdot dx = \int f(x) \cdot dx + \int g(x) \cdot dx$$

This gives us enough machinery to be able to compute the integrals of polynomial functions.

Example 1.3   Find

$$\int (x^3 + 4x^2 + 5x + 3) \cdot dx$$

Solution:

$$\int (x^3 + 4x^2 + 5x + 3) \cdot dx = \int x^3 \cdot dx + 4 \int x^2 \cdot dx + 5 \int x^1 \cdot dx + 3 \int dx$$

$$= \frac{1}{4}x^4 + 4 \cdot \frac{1}{3}x^3 + 5 \cdot \frac{1}{2}x^2 + 3 \cdot x + C$$

$$= \frac{1}{4}x^4 + \frac{4}{3}x^3 + \frac{5}{2}x^2 + 3x + C$$

◊

Notice that an anti-derivative of a constant a is ax and that, since the sum of several unknown constants is some other unknown constant, we can get away with one "+C".

To check an integral rule, you take the derivative of the result and see if you get back where you started.

Example 1.4   Check the preceding example by taking the derivative.

Solution:

$$\left(\frac{1}{4}x^4 + \frac{4}{3}x^3 + \frac{5}{2}x^2 + 3x + C \right)' = \frac{1}{4}4x^3 + \frac{4}{3}3x^2 + \frac{5}{2}2x^1 + 3(1) + 0 = x^3 + 4x^2 + 5x + 3$$

and we see the integral in the last example was correct.

◊

All the derivative rules from *Fast Start Differential Calculus* have corresponding integral rules. Let's go through them.

Knowledge Box 1.5

Integrals of logs and exponentials

$$\int \frac{1}{x} \cdot dx = \ln(x) + C$$

$$\int e^x \cdot dx = e^x + C$$

One problem with the laundry-list of integrals in this section is that, until we get the techniques in Chapter 4, the integrals we can do are a bit contrived. Nevertheless, let's do an example.

Example 1.5   Compute

$$\int \left(3e^x + \frac{4}{x}\right) \cdot dx$$

Solution:

$$\int \left(3e^x + \frac{4}{x}\right) \cdot dx = 3 \int e^x \cdot dx + 4 \int \frac{dx}{x}$$

$$= 3e^x + 4\ln(x) + C$$

◊

Next are the integrals arising from the trigonometric derivatives.

Knowledge Box 1.6

Integrals of trigonometric derivatives

$$\int \sin(x) \cdot dx = -\cos(x) + C$$

$$\int \cos(x) \cdot dx = \sin(x) + C$$

$$\int \sec^2(x) \cdot dx = \tan(x) + C$$

$$\int \csc^2(x) \cdot dx = -\cot(x) + C$$

$$\int \sec(x)\tan(x) \cdot dx = \sec(x) + C$$

$$\int \csc(x)\cot(x) \cdot dx = -\csc(x) + C$$

In the next example, the integral we are asked to perform doesn't look like one of our known forms, but it can be rearranged into one of the known forms with a few trig identities.

Example 1.6   Compute

$$\int \frac{\sin(x)}{\cos^2(x)} \cdot dx$$

Solution:

$$\int \frac{\sin(x)}{\cos^2(x)} \cdot dx = \int \frac{1}{\cos x} \cdot \frac{\sin(x)}{\cos(x)} \cdot dx$$

$$= \int \sec(x) \cdot \tan(x) \cdot dx$$

$$= \sec(x) + C$$

◊

Trigonometric identities can be used to make really easy integrals look really hard.

Example 1.7   Compute

$$\int \left(\sin^2(x) + 2\sin(x) + \cos^2(x) \right) \cdot dx$$

Solution:

$$\int \left(\sin^2(x) + 2\sin(x) + \cos^2(x) \right) \cdot dx = \int \left(\sin^2(x) + \cos^2(x) \right) \cdot dx + 2 \int \sin(x) \cdot dx$$

$$= \int 1 \cdot dx + 2 \int \sin(x) \cdot dx$$

$$= x + 2(-\cos(x)) + C$$

$$= x - 2\cos(x) + C$$

$$\Diamond$$

We also have some integrals arising from the inverse trigonometric functions.

Knowledge Box 1.7

More integrals of trigonometric derivatives

$$\int \frac{1}{\sqrt{1 - x^2}} \cdot dx = \sin^{-1}(x) + C$$

$$\int \frac{1}{1 + x^2} \cdot dx = \tan^{-1}(x) + C$$

$$\int \frac{1}{x\sqrt{x^2 - 1}} \cdot dx = \sec^{-1}(|x|) + C$$

This example is another one in which we can set up a known form with a little bit of algebra.

Example 1.8   Compute

$$\int \left(\frac{1}{3x^2 + 3} \right) \cdot dx$$

Solution:

$$\int \frac{1}{3x^2 + 3} \cdot dx = \int \frac{1}{3}\left(\frac{1}{x^2 + 1}\right) \cdot dx$$

$$= \frac{1}{3}\int \frac{dx}{1 + x^2}$$

$$= \frac{1}{3}\tan^{-1}(x) + C$$

$\diamond$

It is quite common for integrals to require some algebraic setup before it becomes apparent how to do them. Let's do another.

Example 1.9   Compute

$$\int \left(\frac{x^2}{1 + x^2}\right) \cdot dx$$

Solution:

$$\int \frac{x^2}{1 + x^2} \cdot dx = \int \frac{1 + x^2 - 1}{1 + x^2} \cdot dx$$

$$= \int \frac{1 + x^2}{1 + x^2} \cdot dx - \int \frac{dx}{1 + x^2}$$

$$= \int dx - \int \frac{dx}{1 + x^2}$$

$$= x - \tan^{-1}(x) + C$$

$\diamond$

Example 1.10   Compute

$$\int \frac{x\sqrt{1-x^2}+1}{\sqrt{1-x^2}}\,dx$$

Solution:

$$\int \frac{x\sqrt{1-x^2}+1}{\sqrt{1-x^2}}\,dx = \int \left(\frac{x\sqrt{1-x^2}}{\sqrt{1-x^2}} + \frac{1}{\sqrt{1-x^2}}\right)dx$$

$$= \int \left(x + \frac{1}{\sqrt{1-x^2}}\right)dx$$

$$= \frac{1}{2}x^2 + \sin^{-1}(x) + C$$

$$\Diamond$$

All of the integrals we've done in this section lead to formulas, not numbers. That will change in the next section. But before we're done, let's get the terminology for these "formula only" integrals.

Definition 1.3 *The integral formulas given in this section of the form:*

$$\int f(x)\cdot dx = F(x) + C$$

are called **indefinite integrals** *in honor of the unknown constant C.*

One way to think of indefinite integrals is they store patterns that we will use in the integrals that compute specific quantities.

The integrals presented in this section are all simple anti-derivatives, except for the ones that have been lightly disguised by the use of algebra. The next three sections explore the applications and origins of integration without expanding the tool set for actually performing integration. Chapter 4 develops u-substitution, integration by parts, partial fractions, and many other clever methods of doing integrals. It also develops some methods based on trig identities that turn calculus into a puzzle solving activity.

First introducing integration and later presenting the starter tool-kit is something that is done to hand you integration as a useful tool for physics as soon as possible. The later exploration of integration techniques is mind-expanding, but the fundamental concept that permits you to manipulate the formulas that describe natural law is more important in preparing you to study physics.

PROBLEMS

Problem 1.11   Find the indefinite integral of each of the following polynomial functions.

1. $f(x) = x^2 + 1$

2. $g(x) = (x + 1)^2$

3. $h(x) = x^4 - 3x^3 + 5x^2 - 7x + 6$

4. $r(x) = \dfrac{x^3 - 1}{x + 1}$

5. $s(x) = 7x^7 + 6x^6 + x - 2$

6. $q(x) = (x + 1)(x + 2)(x + 3)$

7. $a(x) = (x^2 + x + 1)^3$

8. $b(x) = (x + 2)^5$

Problem 1.12   Perform each of the following integrals. Hint: for many of these, trig identites may be a huge help.

1. $\displaystyle\int \frac{x^4}{x^2 + 1} dx$

2. $\displaystyle\int \frac{x^2 + 3x + 5}{x} dx$

3. $\displaystyle\int \frac{dx}{\cos^2(x)}$

4. $\displaystyle\int \frac{dx}{\sin^2(x) + 4 + 5x^2 + \cos^2(x)}$

5. $\displaystyle\int \left(\frac{1}{\cos^2(x)} + \frac{\sin(x)\sec(x)}{\cos(x)} \right) dx$

6. $\displaystyle\int (\sin^2(x) + \cos^2(x) +$ $\sin(x) + \cos(x) + 1) \cdot dx$

7. $\displaystyle\int 2\sin(x/2)\cos(x/2) \cdot dx$

8. $\displaystyle\int \sin\left(x + \frac{\pi}{3} \right) \cdot dx$

Problem 1.13   For each of the following statements, verify that the integral is correct, by taking a derivative.

1. $\ln(x^2 + 1) + C = \displaystyle\int \frac{2x}{x^2 + 1} \cdot dx$

2. $\dfrac{1}{2} \sin(2x) + C = \displaystyle\int \cos(2x) \cdot dx$

3. $\ln(\sec(x)) + C = \displaystyle\int \tan(x) \cdot dx$

4. $\dfrac{1}{10}(x^2 + 1)^5 + C = \displaystyle\int x \cdot (x^2 + 1)^4 \cdot dx$

5. $\ln(\sec(x) + \tan(x)) + C = \displaystyle\int \sec(x) \cdot dx$

6. $(x - 1)\,e^x + C = \displaystyle\int xe^x \cdot dx$

7. $\dfrac{1}{2}\ln\left(\dfrac{x-3}{x-1}\right) + C = \displaystyle\int \dfrac{dx}{x^2 - 4x + 3}$

8. $(x^3 - 3x^2 + 6x - 6)e^x + C = \displaystyle\int x^3 e^x \cdot dx$

Problem 1.14   Verify that

$$\int \frac{dx}{ax + b} = \frac{1}{a}\ln(ax + b) + C$$

Problem 1.15   Verify that

$$\int e^{ax} = \frac{1}{a}e^{ax} + C$$

Problem 1.16   Verify that

$$\int \cos(ax + b) = \frac{1}{a}\sin(ax + b)$$

Problem 1.17   Verify that

$$\int \sec^2(ax + b) = \frac{1}{a}\tan(ax + b)$$

Problem 1.18   Demonstrate that if $p(x)$ is a polynomial, then we can compute

$$\int \frac{p(x)}{x^k} \cdot dx$$

with the techniques in this section.

Problem 1.19   Compute

$$\int \frac{x^2 + 2}{x^2 + 1} \cdot dx$$

Problem 1.20   Compute

$$\int \frac{\sqrt{1-x^2}}{(1-x)(1+x)} \cdot dx$$

1.2   THE FUNDAMENTAL THEOREM

Before we can formally state the relationship between the integral and the derivative, we need to define the **definite integral**.

Definition 1.4 *Suppose that* $F(x) = \int f(x) \cdot dx$. *In other words,* $F(x)$ *is an anti-derivative of* $f(x)$. *Then the definite integral from* $x = a$ *to* $x = b$ *of* $f(x)$ *is defined to be:*

$$\int_a^b f(x) \cdot dx = F(b) - F(a)$$

One nice thing about the definite integral is that it removes the unknown constant. If we write $F(x) + C$ for the anti-derivative, then

$$(F(b) + C) - (F(a) + C) = F(b) - F(a) + C - C = F(b) - F(a)$$

With the definite integral defined we can now state the first form of the fundamental theorem.

Knowledge Box 1.8

The First Fundamental Theorem of Calculus

$$\frac{d}{dx} \int_a^x f(t) \cdot dt = f(x)$$

for any constant a.

This form of the fundamental theorem tells us that the derivative of the integral of a function is the same function, although the variable of integration (t in the above statement) may be different from the variable appearing in the final expression.

Example 1.21

$$\frac{d}{dx} \int_0^x (t^2 + 1) \cdot dt = x^2 + 1$$

$\Diamond$

The integral of the derivative is also the same function – almost. The ubiquitous unknown constant causes us to answer: except for the "+ C." Later, we will see that this unknown constant is where we place the starting point (position, velocity, etc.) into the formula when solving an applied problem.

The second form of the fundamental theorem has more applications. It relates integrals to the area under the graph of a function.

Knowledge Box 1.9

The Second Fundamental Theorem of Calculus

Suppose that $f(x) \geq 0$ on the interval $[a,b]$. Then, if A is the area under the graph of $f(x)$ between a and b,

$$A = \int_a^b f(x) \cdot dx$$

Example 1.22   Find the area A under the curve of $y = x^2$ between $x = 0$ and $x = 3$.

Solution:

Guided by the picture, compute the definite integral.

Start with a picture:

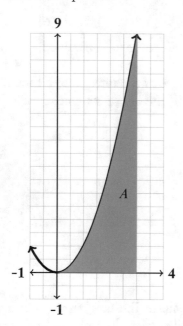

$$\int_0^3 x^2 \cdot dx = \left.\frac{1}{3}x^3\right|_0^3$$

$$= \frac{1}{3}3^3 - \frac{1}{3}0^3$$

$$= \frac{1}{3}27 - \frac{1}{3}0$$

$$= 9 - 0$$

$$= 9 \text{ units}^2$$

◊

Notice the vertical bar notation, used to hold the limits until we plug them into the anti-derivative.

At this point let's check the intuition on this one. Why would the anti-derivative of a function be the area under it? The first form of the fundamental theorem tells us that a function is the derivative of its integral – but that means that a function is the **rate of change** of its integral. The larger a function is, the faster the area under it is changing. The smaller a function is, the slower the area under it is changing. So, a function is the rate of change of the area under the function. If you're unconvinced, wait for Section 1.4. We will use another approach to show that the integral gives the area under the curve.

What meaning does the restriction $f(x) \geq 0$ in the second fundamental theorem have? The short answer is: the area below the x axis, for which $f(x) \leq 0$, comes out negative. This actually makes sense if we remember that the derivative is a rate of change. Positive derivatives represent increases, negative ones represent decreases. Since there is no such thing as negative area, we have to be careful when computing the total area between a graph and the x-axis.

Example 1.23   Compare the definite integral and the area between the curve and the x-axis for $f(x) = x^2 - 1$ from $x = -1$ to $x = 2$.

Solution:

This picture shows the areas above and below the x-axis.

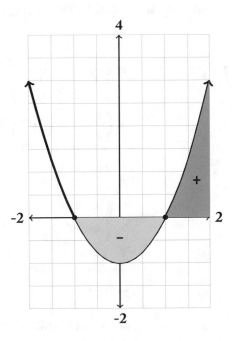

Notice the points where $f(x)$ crosses the axis are $x = \pm 1$.

First the integral:

$$\int_{-1}^{2} (x^2 - 1) \cdot dx = \frac{1}{3}x^3 - x \Big|_{-1}^{2}$$

$$= \frac{1}{3}2^3 - 2 - \left(\frac{1}{3}(-1)^3 - (-1)\right)$$

$$= \frac{8}{3} - 2 + \frac{1}{3} - 1$$

$$= \frac{9}{3} - 3$$

$$= 0$$

So, even though they are different shapes, the areas above and below the curve are equal. Now we need to compute the areas separately and take the positive area minus the "negative" one:

$$\int_{1}^{2} (x^2 - 1) \cdot dx - \int_{-1}^{1} (x^2 - 1) \cdot dx = \left(\frac{1}{3}x^3 - x\right)\Big|_{1}^{2} - \left(\frac{1}{3}x^3 - x\right)\Big|_{-1}^{1}$$

$$= \left(\frac{8}{3} - 2 - \frac{1}{3} + 1\right) - \left(\frac{1}{3} - 1 + \frac{1}{3} - 1\right)$$

$$= \frac{7}{3} - 1 - \frac{2}{3} + 2$$

$$= \frac{8}{3} \text{ units}^2$$

$$\Diamond$$

Notice that the integral is a *number* and so has no units, while the area between the curve and the x-axis has Cartesian units squared as its units. It is very important to keep clear in your mind the context in which you are using an integral. The meaning of the result is different for different procedures.

Also notice that we could have found the total area as

$$A = 2\int_1^2 (x^2 - 1) \cdot dx$$

once we knew the areas above and below the curve were equal. Since it's not obvious until *after* you do the integral, this isn't all that useful in this case. We will look at a useful version of this phenomenon later.

Example 1.24   Find the area under $y = \dfrac{1}{x}$ from $x = 1$ to $x = 4$.

Solution:

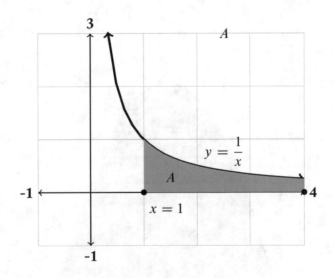

$$\int_1^4 \frac{1}{x} \cdot dx = \ln(x)\,\Big|_1^4$$

$$= \ln(4) - \ln(1)$$

$$= \ln(4) - 0$$

$$= \ln(4)$$

◊

This example pays off on explaining a mystery – why we use e $\cong 2.71828\ldots$ as the base of the "natural" logarithm. It is because the area under $y = \dfrac{1}{x}$ from $x = 1$ to $x = a$ is $\ln(a)$. This gives us a method of computing logs, and it shows a place where logarithms arise naturally from the rest of mathematics. A much better way to choose a base than "we have ten fingers."

Example 1.25 Find the area bounded by $y = 4 - x^2$ and the x-axis.

Solution:

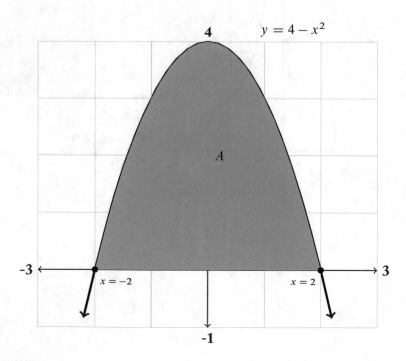

This problem does not give explicit limits – instead it tells us what the objects bounding the area are. Since the x-axis is where $y = 0$, we need to solve $0 = 4 - x^2$, which gives us $x = \pm 2$ as the points at the left and right end of the area. This then gives us a nice everything-above-the-axis integral.

$$A = \int_{-2}^{2} \left(4 - x^2\right) \cdot dx$$

$$= 4x - \frac{1}{3}x^3 \Big|_{-2}^{2}$$

$$= 4(2) - \frac{1}{3}8 - 4(-2) + \frac{1}{3}(-8)$$

$$= 8 + 8 - \frac{8}{3} - \frac{8}{3}$$

$$= 8\left(2 - \frac{2}{3}\right)$$

$$= 8 \cdot \frac{4}{3}$$

$$= \frac{32}{3} \text{ units}^2$$

$$\Diamond$$

Finding the area bounded by two different curves requires solving for those x where

$$curve\ 1 = curve\ 2$$

in order to find the limits of integration. This will come up a lot in the future. This problem can be rephrased as $curve\ 1 - curve\ 2 = 0$ making it a root finding problem.

Example 1.26   Find the area bounded by $y = x^2$ and $y = \sqrt{x}$.

Solution:

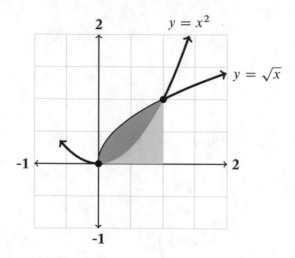

This one is a little tricky. The picture shows us that if we solve $x^2 = \sqrt{x}$ to get the points where the area begins and ends, then those points are (0,0) and (1,1). The area we want is shown in dark gray. It is the area under $y = \sqrt{x}$ that is not under $y = x^2$. If we computed $\int_0^1 \sqrt{x}\,dx$, we would get the dark gray area *and* the light gray area. The light gray area is the area under $y = x^2$. This means the dark gray area is:

$$\int_0^1 \sqrt{x}\,dx - \int_0^1 x^2 \cdot dx$$

Compute

$$\int_0^1 \sqrt{x}\,dx - \int_0^1 x^2 \cdot dx = \int_0^1 \left(x^{1/2} - x^2\right) \cdot dx$$

$$= \frac{2}{3}x^{3/2} - \frac{1}{3}x^3 \Big|_0^1$$

$$= \frac{2}{3} - \frac{1}{3} - 0 + 0$$

$$= \frac{1}{3}\ \text{units}^2$$

$\diamond$

One thing that might be a little tricky in this example is the slightly odd version of the power rule:

$$\int x^{1/2} \cdot dx = \frac{2}{3}x^{3/2} + C$$

All that is going on is that $\frac{1}{2} + 1 = \frac{3}{2}$ and $\frac{1}{3/2} = \frac{2}{3}$.

To find the area bounded by the curves we first found their intersections and then subtracted the area under the lower curve from the area under the upper curve. Another point of view on this is that we integrated the upper curve minus the lower curve. Let's formulate this as a rule.

Knowledge Box 1.10

Area between two curves

If $f(x) > g(x)$ on an interval [a,b], then the area between the graphs of the two functions on that interval is:

$$\int_a^b f(x) \cdot dx - \int_a^b g(x) \cdot dx = \int_a^b (f(x) - g(x)) \cdot dx$$

This rule is handy, but it does leave you with the problem of finding the appropriate intervals.

Example 1.27   Find the area between $y = x^2$ and $y = x + 2$.

Solution:

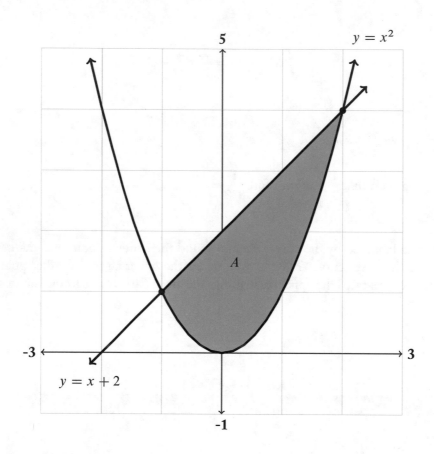

Solving

$$x^2 = x + 2,$$

or equivalently

$$x^2 - x - 2 = 0,$$

we see that $(x - 2)(x + 1) = 0$. So $x = -1, 2$, making the intersection points $(-1, 1)$ and $(2, 4)$. This also gives us the limits of integration and so

$$A = \int_{-1}^{2} \left(x + 2 - x^2 \right) \cdot dt$$

$$= \int_{-1}^{2} \left(2 + x - x^2 \right) \cdot dt$$

$$= 2x + \frac{1}{2}x^2 - \frac{1}{3}x^3 \Big|_{-1}^{2}$$

$$= 4 + \frac{4}{2} - \frac{8}{3} + 2 - \frac{1}{2} - \frac{1}{3}$$

$$= \frac{24 + 12 - 16 + 12 - 3 - 2}{6}$$

$$= 27/6$$

$$= 9/2 \text{ units}^2$$

$$\Diamond$$

Let's do an example that moves beyond polynomial functions.

Example 1.28   Find the area under the sine function from $x = 0$ to $x = \pi$.

Solution:

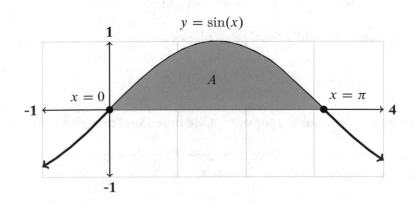

The sine function is non-negative on the entire interval so we may compute

$$A = \int_0^\pi \sin(x) \cdot dx$$

$$= -\cos(x) \Big|_0^\pi$$

$$= -\cos(\pi) + \cos(0)$$

$$= -(-1) + 1$$

$$= 2 \text{ units}^2$$

$$\diamondsuit$$

1.2.1   EVEN AND ODD FUNCTIONS

Even and odd functions have some useful properties, relative to integration. These definitions appear in the book on the differential calculus, but it is worth repeating them here.

Definition 1.5 *A function is* **even** *if, for x where the function exists,*

$$f(x) = f(-x).$$

A good example of an even function is $f(x) = x^2$. **Even functions forget signs.**

Definition 1.6 *A function is* **odd** *if, for x where the function exists,*

$$f(-x) = -f(x).$$

A good example of an odd function is $f(x) = x^3$. **Odd functions remember signs.**

It turns out that we can save some effort when integrating these functions, sometimes, because of special geometric properties of these functions.

Knowledge Box 1.11

Integrating an even function on a symmetric interval

If $f(x)$ is an even function, then

$$\int_{-a}^{a} f(x) \cdot dx = 2 \int_{0}^{a} f(x) \cdot dx$$

Example 1.29   Find the integral of $f(x) = x^2$ on $[-2, 2]$.

Solution:

Remember that $x^2 = (-x)^2$. So $f(x)$ is an even function.

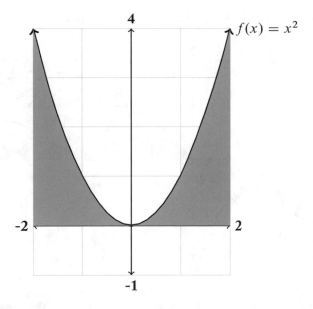

The area on either side of the y-axis is the same so:

$$\int_{-2}^{2} x^2 \cdot dx = 2 \int_{0}^{2} x^2 \cdot dx = 2 \cdot \frac{1}{3} x^3 \Big|_{0}^{2} = \frac{16}{3} \text{ units}^2$$

Not having to plug in the negative number avoids chances to make arithmetic mistakes.

◇

Knowledge Box 1.12

Integrating an odd function on a symmetric interval

If $f(x)$ is an odd function, then

$$\int_{-a}^{a} f(x) \cdot dx = 0$$

Example 1.30   Find the integral of $g(x) = \sin(x)$ on $[-2, 2]$.

Solution:

Remember that $\sin(x) = -\sin(-x)$. So $g(x)$ is an odd function.

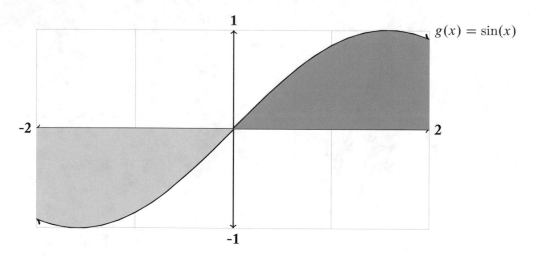

The area on either side of the y-axis is the same, but half is above the x axis, and half is below the x-axis. So:

$$\int_{-2}^{2} \sin(x) \cdot dx = 0$$

Wow, is that easier than plugging in numbers.

◇

The odd and even function results are fairly special purpose. One of the big applications of odd and even function is writing test questions that can be done much faster if the student notices something is an odd function on a symmetric interval, for example. Symmetric intervals *do* also show up in some application problems. If you're good at shifting functions sideways, you can also extend the application of these rules.

The set of even functions and the set of odd functions are closed under addition and also under multiplication by a constant. This fact is occasionally useful. If you're feeling ambitious, try and prove it's true. Also remember that you can break up an integral into a symmetric piece and leftovers, to change a definite integral into a possibly simpler integral. Finally, if you see an integral that you have *no idea* how to do, consider the possibility that the odd function shortcut gives you a way of solving the problem without integrating.

PROBLEMS

Problem 1.31   Simplify each of the following. Assume Snerp(x) has a domain of $(-\infty, \infty)$.

1. $\dfrac{d}{dx} \displaystyle\int_1^x \dfrac{t}{t^2 + 1} \cdot dt$

2. $\dfrac{d}{dx} \displaystyle\int_\pi^x \cos(6t + 1) \cdot dt$

3. $\dfrac{d}{dx} \displaystyle\int_0^x te^{t^2} \cdot dt$

4. $\dfrac{d}{dx} \displaystyle\int_{-6}^x \dfrac{e^s + 1}{e^s - 1} \cdot ds$

5. $\dfrac{d}{dx} \displaystyle\int_2^x \dfrac{e^y}{\ln(y) + 4} \cdot dy$

6. $\dfrac{d}{dx} \displaystyle\int_0^x \text{Snerp}(t + 2) \cdot dt$

Problem 1.32   Compute the following.

1. $\displaystyle\int_{-1}^1 (x^2 + x + 1) \cdot dx$

2. $\displaystyle\int_{-\pi/2}^{\pi/2} \cos(x) \cdot dx$

3. $\displaystyle\int_1^3 \dfrac{1}{x}(x^2 + 1) \cdot dx$

4. $\displaystyle\int_0^4 x^n \cdot dx$

5. $\displaystyle\int_0^1 (x + e^x - 2) \cdot dx$

6. $\displaystyle\int_0^{\sqrt{3}} \dfrac{dx}{x^2 + 1}$

Problem 1.33   Compute

$$\int_{-5}^{5} (x^5 + 3x^3 + 7x^2 - x + 1) \cdot dx$$

Problem 1.34   Compute the area between the following curves and the x axis on the stated interval. Be careful, some of these have area above and below the x-axis. Pictures may help.

1. $f(x) = x^2$ on $[-1, 1]$

2. $g(x) = x^3$ on $[-1, 1]$

3. $h(x) = \cos(x)$ on $[-\pi/2, 3\pi/2]$

4. $r(x) = \tan^{-1}(x)$ on $[-1/\sqrt{3}, 1/\sqrt{3}]$

5. $s(x) = e^{-x}$ on $[0, 5]$

6. $q(x) = 4 - x^2$ on $[-3, 3]$

Problem 1.35   Compute the area bounded by the specified curves. You will need to know which curve is higher. Again: pictures may help.

1. $y = x^2$ and $y = x^3$

2. $y = x^2$ and $y = 6 - x$

3. $y = \sin(x)$ and $y = \cos(x)$ on $[0, 3\pi/2]$

4. $y = x^2$ and $y = 4x - 3$

5. $y = x^2$ and $y = 9 - x^2$

6. $y = x^3$ and $y = 4x$

Problem 1.36   Find the area bounded by

$$y = x^n \text{ and } y = x^m$$

for all positive whole numbers $m < n$. There will be different categories of answer based on whether n and m are even or odd. That's four categories: $++$, $+-$, $-+$, and $--$.

Problem 1.37   Compute the following definite integrals.

1. $\int_{-3}^{3} (x^5 + x^3 + x + 1) \cdot dx$

2. $\int_{0}^{2\pi} (\cos(x) + \sin(x)) \cdot dx$

3. $\displaystyle\int_{-1}^{1} (x^6 + x^5 + x^4 + x^3 + x^2 + x + 1) \cdot dx$

4. $\displaystyle\int_{-3\sqrt{3}}^{3\sqrt{3}} \tan^{-1}(x) \cdot dx$

5. $\displaystyle\int_{0}^{2\pi} (\sin(x) + \sin(2x) * + \sin(3x)) \cdot dx$

6. $\displaystyle\int_{-a}^{a} x^{2n} \cdot dx$

7. $\displaystyle\int_{-a}^{a} x^{2n+1} \cdot dx$

8. $\displaystyle\int_{-a}^{a} (\sin(x) + tan^{-1}(x) + x^5) \cdot dx$

Problem 1.38   Classify each of the following functions as being odd, even, or neither.

1. $y = \sin(x)$

2. $y = \cos(x)$

3. $y = \tan^{-1}(x)$

4. $y = \ln(x^2 + 1)$

5. $y = x \cdot \ln(x^2 + 1)$

6. $y = e^{-x^2/2}$

7. $y = x^2 + x + 1$

8. $y = \sin(x^2)$, and

9. $p(x^2)$, where $p(x)$ is any polynomial.

Problem 1.39   Suppose that $f(x)$ is an even function. Prove that $y = x \cdot f(x)$ is an odd function.

Problem 1.40   Suppose that $f(x)$ is an odd function. Prove that $y = x \cdot f(x)$ is an even function.

Problem 1.41   Suppose that $f(x)$ is a function and $g(x)$ is an even function. Is $f(g(x))$ an even function? Demonstrate your answer is correct.

Problem 1.42   Find b when

$$\int_{0}^{b} x^2 \cdot dx = 14$$

Problem 1.43   Find b when

$$\int_0^b x^4 \cdot dx = 1$$

Problem 1.44   If

$$\int_0^\theta \sin(x) \cdot dx = 2,$$

what is the smallest possible value for θ?

Problem 1.45   Find a constant c so that the area bounded by $y = c$ and $y = x^2$ is exactly 4 units^2.

Problem 1.46   Compute the slope m so that the area bounded by the curves $y = x^3$ and $y = mx$ is exactly 4 units.

Problem 1.47   Find the largest possible value of $\int_a^b \cos(x) \cdot dx$.

Problem 1.48   Compute

$$\int_{-5}^5 x^3 \cdot \ln\left(x^2 + 1\right)\ dx$$

Problem 1.49   Suppose

$$F(x) = \int_0^x e^{t^2} \cdot dt$$

What is $F'(x)$?

1.3   INITIAL VALUE PROBLEMS

In this section we come to grips with the constant of integration and figure out what its value is. This requires that we have a bit of additional information. Our motivating example is to build up the position function $s(t)$ from the acceleration function $a(t)$ in steps, with the velocity function $v(t)$ as an intermediate object. The mathematical model of motion in one dimension under constant acceleration is:

$$s(t) = \frac{1}{2}at^2 + v_0t + s_0$$

In English, the distance an object is from a reference point is equal to half the acceleration times the time squared plus the initial velocity times the time plus the initial distance from that reference point. If we break this into integrals we get:

$$v(t) = \int_{t_0}^{t} a \cdot dx$$

$$= a \cdot (t - t_0) + C$$

$$v(t_0) = a \cdot 0 + C$$

$$v_0 = C$$

So the constant of integration when we transform acceleration into velocity is the initial velocity. Similarly:

$$s(t) = \int_{t_0}^{t} v \cdot dx$$

$$= v \cdot (t - t_0) + C$$

$$s(t_0) = v \cdot 0 + C$$

$$s_0 = C$$

The constant of integration for velocity is initial distance. This shows how the constant of integration can be solved for if we know the initial value of the quantity we are calculating. In fact, all we need is the value of the thing we are calculating *anywhere* in the interval we are integrating on. The initial value just has neater algebra.

Example 1.50   Suppose we fire a cannonball directly upward with a velocity of 120 m/sec with a gravitational acceleration of 10 m/sec^2. If the cannon is at a height of 160 m above sea level, find an expression for the distance above sea level of the cannonball after the cannon is fired at $t = 0$.

Solution:

Plug into the equation of motion given above.

$$s(t) = -\frac{1}{2}10t^2 + 120t + 160 = 160 + 120t - 5t^2$$

Since we have a model for this situation, the calculus is all done. Let's look at a situation where we need calculus.

◇

Example 1.51   Suppose a missile has an acceleration that builds gradually so that $a(t) = 100 + 0.1t$. If it is launched from a fixed position with a charge that gives it an initial velocity of 20m/sec, find an expression for the distance the missile has traveled t seconds after launch and find its position and velocity after $t = 20$ sec when its fuel runs out.

Solution:

First we find the velocity function. Assume that we start at $t = 0$.

$$v(t) = \int a(t) \cdot dt$$

$$= \int (100 + 0.1t) \cdot dt$$

$$= 100t + 0.05t^2 + C \cdot$$

Now solve for C:

$$20 = v(0)$$

$$20 = 100(0) + 0.05(0) + C$$

$$20 = C$$

And we obtain:

$$v(t) = 0.05t^2 + 100t + 20$$

This gives us the velocity after $t = 20$ sec: $V(20) = 2040$ m/sec. Now we find the distance function.

$$s(t) = \int v(t) \cdot dt$$

$$= \int (0.05t^2 + 100t + 20) \cdot dt$$

$$= \frac{1}{60}t^3 + 50t^2 + 20t + C$$

As before...

$$s(0) = 0 + 0 + 0 + C = 0$$

$$C = 0$$

So...

$$s(t) = \frac{1}{60}t^3 + 50t^2 + 20t$$

This is the expression for the the distance the missile has traveled t seconds after launch, and its position after $t = 20$ sec is $s(20) \cong 20,533$ m.

◊

Example 1.52   Suppose that

$$f(x) = \int (3x^2 + 1)\, dx$$

and we know that $f(2) = 3$. Find an expression for $f(x)$ with no unknown constants.

Solution:

$$f(x) = \int (3x^2 + 1)\, dx$$

$$= x^3 + x + C$$

Now use the added information.

$$f(2) = 2^3 + 2 + C$$

$$3 = 10 + C$$

$$-7 = C$$

Combine

$$f(x) = x^3 + x - 7$$

$$\Diamond$$

Knowledge Box 1.13

Solving initial value problems

When a function resulting from integration has the form

$$f(x) + C,$$

an additional piece of information is needed to determine a value for C.

Example 1.53   Suppose that $g(x) = \int e^x \cdot dx$. For $g(1) = 12.2$, find an expression for $g(x)$ that does not involve any unknown constants.

Solution:

The integral is trivial – e^x is its own integral – and so

$$g(x) = e^x + C$$

Plug in the additional information and solve.

$$g(1) = 12.2$$

$$e^1 + C = 12.2$$

$$C = 12.2 - e$$

$$C \cong 9.48$$

$$\Diamond$$

If we need to do more than one integral, we will need one piece of added information per unknown constant that arises.

Example 1.54   Suppose that the second derivative of $h(x)$ is $h''(x) = 1.2x - 1$. For $h(0) = 4$ and $h'(2) = 2$, find an expression for $h(x)$ that is free of unknown constants.

First we find the first derivative of $h(x)$:

$$h'(x) = \int (1.2\,x - 1)dx$$

$$= 0.6x^2 - x + C_1$$

Plug in the value of the derivative:

$$2 = h'(2)$$

$$= 0.6(2)^2 - 2 + C_1$$

$$= 0.4 + C_1$$

$$1.6 = C_1$$

$$h'(x) = 0.6x^2 - x + 1.6$$

Now we move on to the function itself:

$$h(x) = \int (0.6x^2 - x + 1.6)dx$$

$$= 0.2x^3 - 0.5x^2 + 1.6x + C_2$$

$$4 = h(0)$$

$$4 = 0 + 0 + 0 + C$$

$$4 = C$$

$$h(x) = 0.2x^3 - 0.5x^2 + 1.6x + 4$$

And we have our answer. Notice that, since two unknown constants appeared, we gave them different names: C_1 and C_2. Also notice that it is much easier to deal with added information or initial conditions that happen at time zero.

$$\diamond$$

PROBLEMS

Problem 1.55   Assume we are describing the upward motion of a projectile. Given the constant acceleration, initial velocity, and initial position, determine the greatest height of the projectile and the time it has that height.

1. $a = -5 \text{ m/s}^2$, $v_0 = 20$ m/sec, and $s_0 = 5$ m

2. $a = -5 \text{ m/s}^2$, $v_0 = 10$ m/sec, and $s_0 = -4$ m

3. $a = -5 \text{ m/s}^2$, $v_0 = 12$ m/sec, and $s_0 = 3$ m

4. $a = -8 \text{ m/s}^2$, $v_0 = 30$ m/sec, and $s_0 = -4.1$ m

5. $a = -2 \text{ m/s}^2$, $v_0 = 11$ m/sec, and $s_0 = 1.4$ m

6. $a = -14 \text{ m/s}^2$, $v_0 = 14.2$ m/scc, and $s_0 = 7.33$ m

Problem 1.56   Find $s(t)$ given the following rates of acceleration and initial velocity and position.

1. $a(t) = 3.2 + 1.1 \cdot t \text{ m/s}^2$, $v_0 = 3$ m/s, $s_0 = 0$ m

2. $a(t) = \sqrt{t} \text{ m/s}^2$, $v_0 = -1$ m/s, $s_0 = 5$ m

3. $a(t) = t^{1.5} \text{ m/s}^2$, $v_0 = 4$ m/s, $s_0 = -2$ m

4. $a(t) = t^2 - t \text{ m/s}^2$, $v_0 = 2$ m/s, $s_0 = 0.4$ m

5. $a(t) = \cos(t) \text{ m/s}^2$, $v_0 = 1$ m/s, $s_0 = 0$ m

6. $a(t) = \dfrac{1}{t} \text{ m/s}^2$, $v_0 = -1$ m/s, $s_0 = 0$ m

Problem 1.57   Solve the following initial value problems.

1. $f(x) = \displaystyle\int x^2 - x \cdot dx$ when $f(0) = 3$

2. $g(x) = \displaystyle\int -\sin(x) \cdot dx$ when $g(0) = 0.7071$

3. $h(x) = \displaystyle\int (0.04x^3 - 2x + 4) \cdot dx$ when $h(1) = 2.4$

4. $r(x) = \displaystyle\int \left(0.08x + \frac{2.3}{x} \right) \cdot dx$ when $r(2) = 0.4$

5. $s(x) = \displaystyle\int \frac{dx}{x^2 + 1}$ when $s(1) = 1.2$

6. $q(x) = \displaystyle\int \frac{x^2 + 1}{x} \cdot dx$ when $q(1) = 8.2$

Problem 1.58   Suppose that
$$f''(x) = 2x + 1$$
If $f(1) = 3$ and $f'(1) = 0.5$, find an expression for $f(x)$ with no unknown constants.

Problem 1.59   Suppose that
$$g''(x) = 0.03x^2 - 4x + 1$$
If $g(2) = 7$ and $g'(1) = 3$, find an expression for $g(x)$ with no unknown constants.

Problem 1.60   Suppose that
$$h''(x) = 1.4x + 5.6$$
If $h(1) = 2$ and $h(2) = 4$, find an expression for $h(x)$ with no unknown constants. Notice that you will have to integrate an unknown constant before you can solve.

Problem 1.61   Suppose that
$$r''(x) = 2e^x$$
If $r(0) = 2$ and $r'(0) = 8$, find an expression for $r(x)$ with no unknown constants.

Problem 1.62   Suppose that
$$s''(x) = e^x$$
If $s(0) = 1.2$ and $s(1) = 5$, find an expression for $s(x)$ with no unknown constants. Notice that you will have to integrate an unknown constant before you can solve.

Problem 1.63   Suppose that
$$q''(x) = \sin(x) + \cos(x)$$
If $q(0) = 1.0$ and $q(\pi/2) = 0.5$, find an expression for $q(x)$ with no unknown constants. Notice that you will have to integrate an unknown constant before you can solve.

1.4   INDUCTION AND SUMS OF RECTANGLES

In this section we will study the theory of integration that was developed before the fundamental theorem that doesn't need the concept of the derivative. Let's start with an example that could be solved with calculus but need not be.

Example 1.64   Find, without using calculus, the area under $f(x) = 2x + 1$ from $x = 0$ to $x = 3$.

Solution:

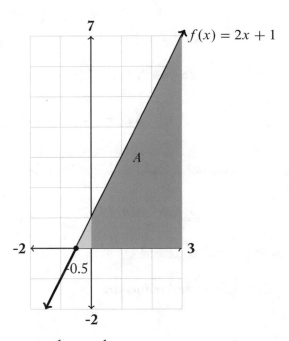

The area of the large triangle is $\frac{1}{2}bh = \frac{1}{2}3.5 \times 7 = 12.25$. The area of the light gray triangle is also $\frac{1}{2}bh = \frac{1}{2}0.5 \times 1 = 0.25$. The area we want, the dark gray, is thus $12.25 - 0.25 = 12$ units^2.

◊

The reason we don't need calculus for this example is that the shape of the area is really simple. As long as we remember that the area of a triangle is one-half base times height, we're fine. As soon as the integral is under a function that is curved, we will need calculus. Where does this calculus come from? At this point we need a digression to create the mathematical machinery that permits us to synthesize the calculus.

The symbol

$$\sum$$

means "add up the things following the symbol." So

$$\sum_{i=1}^{5} i$$

is a symbolic way of saying "add up the numbers i from $i = 1$ to $i = 5$. The name of the symbol is **uppercase sigma** – not to be confused with σ which is called **lowercase sigma**. Applying $\sum$ we get

$$\sum_{i=1}^{5} i = 1 + 2 + 3 + 4 + 5 = 15$$

which isn't too bad. The problem is when we get something like

$$\sum_{i=1}^{200} i = 1 + 2 + \cdots + 199 + 200 = 20,100$$

The incomparable German mathematician Carl Friedrich Gauss found the following shortcut:

$$\sum_{i=1}^{n} i = \frac{1}{2}n(n + 1)$$

How do you prove a formula like that is correct?

The usual technique is called **mathematical induction**.

Knowledge Box 1.14

Mathematical induction

Suppose we wish to prove a proposition $P(n)$ is correct for all $n \geq c$. Then the following steps suffice

- *Verify that $P(c)$ is true.*

- *Assume that, for some n, $P(n)$ is true.*

- *Show that, if $P(n)$ is true, then so is $P(n + 1)$*

The assumption in the second step is easy, because $P(n)$ is true when $n = c$, due to our work on the first step. The key step is the third one. Once we've got it, $P(c)$ implies $P(c + 1)$ is true, which in turn implies $P(c + 2)$ is true, and so on, until we hit any particular $n \geq c$.

Example 1.65   Use mathematical induction to prove Gauss' formula.

Solution:

The proposition is

$$P(n) \ : \ \sum_{i=1}^{n} i = \frac{1}{2}n(n + 1)$$

Let's start with $c = 1$.

$$P(c) = P(1) : \sum_{i=1}^{1} i = 1 = \frac{1}{2}c(c + 1) = \frac{1}{2}(1)(1 + 1) = 1$$

Since 1=1, this is true, and we have the first step of the induction. We now assume that $P(n)$ is true (for some n) and look at $P(n + 1)$.

$$\sum_{i=1}^{n+1} i = \sum_{i=1}^{n} i + (n + 1)$$

$$= \frac{1}{2}n(n + 1) + (n + 1) \qquad\qquad \text{Apply } P(n)$$

$$= \frac{n(n + 1) + 2(n + 1)}{2}$$

$$= \frac{n^2 + 3n + 2}{2}$$

$$= \frac{1}{2}(n + 1)(n + 2)$$

$$= \frac{1}{2}(n + 1)((n + 1) + 1) \qquad\qquad \text{Which is } P(n + 1)$$

So if $P(n)$ is true, we can show using algebra that $P(n + 1)$ is true. This tells us, by mathematical induction, that Gauss' formula is true for all $n \geq 1$.

$\diamond$

Some other examples of proof with mathematical induction appear in the homework problems. Now we know enough to calculate an integral or a curved function *without* using the fundamental theorem of calculus. Our goal is to approximate the area under the curve with shapes we can compute.

Examine pictures of $f(x) = x^2$ in Figure 1.1. The gray area is $\int_0^2 x^2 \cdot dx$; the four rectangles are a (bad) approximation of the area under the curve. How can we make the approximation better? Use more rectangles! The more rectangles, the better the area is approximated.

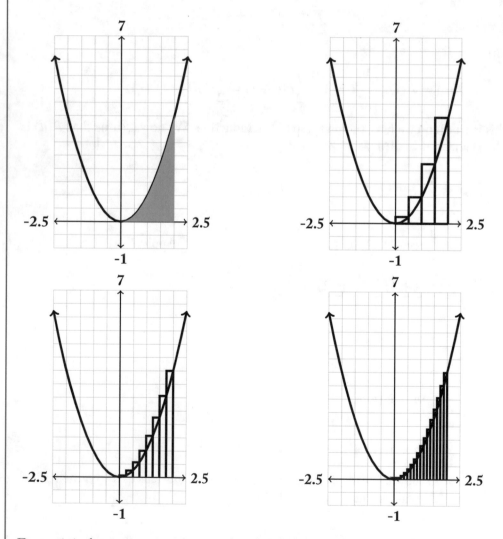

Figure 1.1: Approximating an intgral with increasing number of rectangles.

It turns out that summing and integrating share two algebraic properties.

Knowledge Box 1.15

Algebraic properties of $\sum$

$$\sum_{i=a}^{b} f(i) + g(i) = \sum_{i=a}^{b} f(i) + \sum_{i=a}^{b} g(i)$$

$$\sum_{i=a}^{b} c \cdot f(i) = c \cdot \sum_{i=a}^{b} f(i)$$

for any constant c

There are an infinite number of summation formulas – but a few of them are especially useful. All of these can be proved by induction (and you are asked to do so in the homework).

Knowledge Box 1.16

Useful sum formulas

$$\cdot \ \sum_{i=1}^{n} 1 = n$$

$$\cdot \ \sum_{i=1}^{n} i = \frac{n(n+1)}{2}$$

$$\cdot \ \sum_{i=1}^{n} i^2 = \frac{n(n+1)(2n+1)}{6}$$

$$\cdot \ \sum_{i=1}^{n} i^3 = \frac{n^2(n+1)^2}{4}$$

If we use n rectangles to approximate $\int_{1}^{2} x^2 \cdot dx$, then each rectangle has a width of $w = \frac{2}{n}$. The right side of each rectangle (which determines its height) is $x_i = i \cdot w$ for $i = 1, 2, \ldots n$.

The height of the ith rectangle is $(x_i)^2 = (iw)^2 = \left(\dfrac{2i}{n}\right)^2$. Summing the areas we get that:

$$A \cong \sum_{i=1}^{n} W \times H = \sum_{i=1}^{n} \frac{2}{n}\left(\frac{2i}{n}\right)^2$$

Let's simplify this with the algebraic rules for sums.

$$\sum_{i=1}^{n} \frac{2}{n}\left(\frac{2i}{n}\right)^2 = \sum_{i=1}^{n} \frac{2}{n}\frac{4i^2}{n^2}$$

$$= \sum_{i=1}^{n} \frac{8i^2}{n^3}$$

$$= \frac{8}{n^3}\sum_{i=1}^{n} i^2$$

$$= \frac{8}{n^3}\frac{n(n+1)(2n+1)}{6} \qquad\qquad \text{Use } \sum i^2 \text{ formula}$$

$$= \frac{16n^3 + 24n^2 + 8n}{6n^3}$$

$$= \frac{8n^3 + 12n^2 + 4n}{3n^3}$$

Now we have a formula for the approximate area with n rectangles – the approximation gets better as n grows. This means that

$$\int_0^2 x^2 \cdot dx = \lim_{n\to\infty} \frac{8n^3 + 12n^2 + 4n}{3n^3} = \frac{8}{3} \text{ units}^2$$

which is the same result we get if we do the integral in the usual way.

This is a **very cumbersome** method of computing integrals – not used in practice – but it shows that there is a theory for integrals, just as there is for derivatives. The fundamental theorem is a godsend. Imagine if you did not know that integrals and anti-derivatives were the same thing. Every integral would be a limit of sums of rectangles (or some other shape).

Many integrals cannot be done symbolically – with formulas and algebra. The discipline of **numerical analysis** studies how to use things like rectangle-sum approximations to get useful values for integrals that cannot be calculated with pure calculus.

PROBLEMS

Problem 1.66   Compute the following sums. You may use the formulas you are asked to prove in Problem 1.67.

1. $\displaystyle\sum_{i=1}^{40} i^2$

2. $\displaystyle\sum_{i=20}^{60} i$

3. $\displaystyle\sum_{i=1}^{30} (2i + 5)$

4. $\displaystyle\sum_{i=1}^{100} \frac{1}{2}(i + 2)$

5. $\displaystyle\sum_{i=61}^{101} i$

6. $\displaystyle\sum_{i=5}^{23} 2^i$

7. $\displaystyle\sum_{i=14}^{28} i^3$

8. $\displaystyle\sum_{i=18}^{37} (2i - 1)$

Problem 1.67   Use mathematical induction to demonstrate that the following formulas are correct.

1. $\displaystyle\sum_{i=1}^{n} (2i - 1) = n^2$

2. $\displaystyle\sum_{i=1}^{n} 1 = n$

3. $\displaystyle\sum_{i=1}^{n} i^2 = \frac{n(n + 1)(2n + 1)}{6}$

4. $\displaystyle\sum_{i=1}^{n} i^3 = \frac{n^2(n + 1)^2}{4}$

5. $\displaystyle\sum_{i=0}^{n} 2^i = 2^{n+1} - 1$

6. $\displaystyle\sum_{i=0}^{n} 3^i = \frac{1}{2}\left(3^{n+1} - 1\right)$

7. $\displaystyle\sum_{i=0}^{n} x^i = \frac{x^{n+1} - 1}{x - 1}$

Problem 1.68   Explain why the area under a line $y = mx + b$ can always be found without calculus.

Problem 1.69   A formula for approximating

$$\int_0^2 x^2 \cdot dx$$

with n rectangles was computed in this section. For $n = 4, 6, 8, 12, 20$, and 50 rectangles, compute the error of the approximation.

Problem 1.70   Find the formula for the sum of rectangles for $y = x^3$ to approximate the integral of $y = x^3$ from $x = 0$ to $x = c$. Having found the formula, find the integral by taking a limit.

Problem 1.71   The sum of rectangles used in this section was based on the right side of the intervals. How would an approximation that used the left side of the interval be different? Could it still be used with a limit, to compute integrals? Explain.

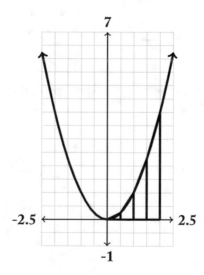

Problem 1.72   The approximation sketched above uses trapezoids instead of rectangles to approximate the area under the curve. The area of each trapezoid is the average of the height of the points on either side of the interval. Write out the approximation formula for n trapezoids instead of n rectangles. Make it as simple as you can.

Problem 1.73   Is the trapezoid method more accurate than the rectangle method? Explain or justify your answer.

Problem 1.74   Approximate

$$\frac{0}{2}x^2 \cdot dx$$

using 20 intervals using rectangles and trapezoids. The area of a trapezoid with left side of height h and right side of height k and width w is

$$A = \frac{1}{2}w(h + k)$$

CHAPTER 2

Parametric, Polar, and Vector Functions

Consider the plot in Figure 2.1. By the rules we've come up with so far this is not even close to being a function. It also looks like the graph of nothing we have done so far. It is a spiral. In this chapter we will look at methods for using calculus – which wants things to act like functions at least locally – on curves like this. The underlying idea is to treat the position of a point as having coordinates that are individually functions of a **parameter**. These are called **parametric functions**.

2.1 PARAMETRIC FUNCTIONS

The spiral shown in Figure 2.1 is not a function of the form $y = f(x)$. It violates the vertical line test and, if we had shown the entire spiral instead of only its beginning, would have intersected every vertical line an infinite number of times. In order to make the spiral – and many other

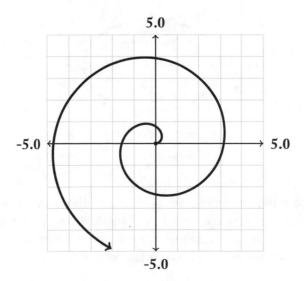

Figure 2.1: Example of a parametric function – a spiral.

curves – into a function, we will make the x and y coordinates of the curve functions of a parameter t in their own right. This is how parametric curves are created.

Knowledge Box 2.1

Parametric curves

If we specify a set of points by

$$(x(t), \; y(t)),$$

where $x = x(t)$ and $y = y(t)$, then the resulting structure is called a **parametric curve**. *The variable t is called the* **parameter**.

Example 2.1   Graph the curve $(x(t), y(t))$ if $x(t) = 3\cos(t)$ and $y(t) = 2\sin(t)$.

Solution:

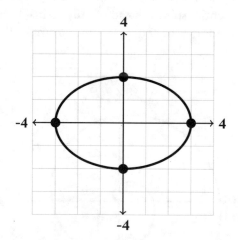

In order to graph this curve, assuming we don't know ahead of time that it is an ellipse with major axis of length 6 and minor axis of length 4, we plot points. We know that $-1 \leq \sin(x), \cos(x) \leq 1$. So, it's not hard to figure out where the bounds are. The points on the curve where the extreme values are have been plotted with dots.

$\Diamond$

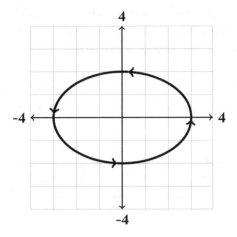

Figure 2.2: Orientation of ellipse from Example 2.1.

Definition 2.1 *The **major axis** of an ellipse is its largest diameter. The **minor axis** of an ellipse is its smallest diameter.*

The major axis of the ellipse in Example 2.1 is 6; the minor axis is 4. In general,

$$(a \cdot \cos(t), b \cdot \sin(t))$$

is an ellipse with major and minor axes in the direction of the coordinate axes, centered at the origin. The major and minor axes have size $2a$ and $2b$ with the major/minor order determined by which is larger.

One thing that is very different about a parametric curve is that the points are **ordered** by the parameter. Points generated by a larger value of the parameter are thought of as coming *after* points generated by a smaller value of the parameter.

Definition 2.2 *The **positive orientation** of a parametric curve is the direction, along the curve, in which the parameter increases.*

The orientation of the ellipse in Example 2.1, shown in Figure 2.2, is counterclockwise.

It turns out that any function of the usual sort can be put into the form of a parametric curve by the simple technique of starting with $y = f(x)$ and defining $x(t) = t$ and $y(t) = f(t)$. This will make the parametric function exactly trace the graph of $y = f(x)$.

One disadvantage of parametric functions is that they give us many, many ways to specify the same function. The lines $(t, 2t + 1)$ and $(3t + 1, 6t + 3)$ are two different parametric forms of the line $y = 2x + 1$.

If $x(t)$ and $y(t)$ are both linear functions, then $(x(t), y(t))$ is the parametric form of *some* line.

Example 2.2   Put the line

$$(2t + 5, 4t - 1)$$

into the standard $y = mx + b$ form.

Solution:

$$x = 2t + 5$$

$$2x = 4t + 10$$

$$2x - 11 = 4t - 1$$

$$2x - 11 = y$$

So we see $y = 2x - 11$ is the standard form of this parametric line.

A big advantage of parametric curves is that they let us specify things that are hard to specify in other ways and that they let us turn things that were not functions into functions of the parameter. Once something is a function, most of the formalisms of calculus apply and we can differentiate, integrate, and optimize.

Example 2.3 Consider the parametric curve $(\sin(t), t\cos(t))$. Graph the curve for $0 \le t \le 2\pi$.

Solution:

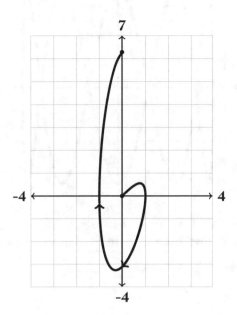

This one is done by plotting points.

$\diamond$

Parametric curves can intersect themselves, possibly many times, at least in the $x - y$ plane. If we consider the parameter, the "intersecting" points are not intersections, because their values of the parameter differ. If we graph $x(t) = 1.8\cos(\frac{4}{3}\pi t)$, $y(t) = 1.8\sin(\frac{20}{3}\pi t)$, we get the curve in Figure 2.3.

We have not yet explained how to make a spiral. If $f(t)$ is a function that is increasing on $[0, \infty)$, then the parametric curve

$$(f(t)\cos(t), f(t)\sin(t))$$

is a spiral with $f(t)$ controlling how fast it spirals outward. If we take $f(t) = \ln(t + 1)$, for $t \ge 0$, then we get the curve in Figure 2.4.

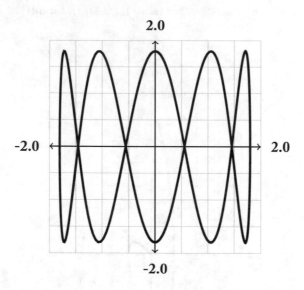

Figure 2.3: Parametric curve that intersects itself. Given by $x(t) = 1.8\cos(\frac{4}{3}\pi t)$, $y(t) = 1.8\sin(\frac{20}{3}\pi t)$.

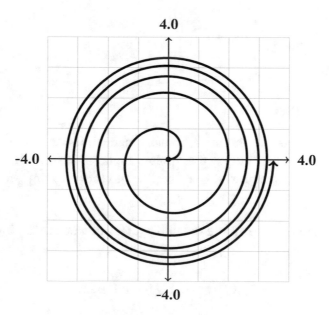

Figure 2.4: Spiral parametric curve. Given by $x(t) = t\cos(t)$, $y(t) = t\sin(t)$

2.1.1 THE DERIVATIVE OF A PARAMETRIC CURVE

In our study of differentials, we learned that we could algebraically cancel differentials. The ordinary derivative of a curve is $\dfrac{dy}{dx}$, which lets us calculate the derivative of a parameteric curve by doing this:

$$\frac{dy}{dx} = \frac{dy/dt}{dx/dt} = \frac{y'(t)}{x'(t)}$$

Let's highlight that in a Knowledge Box.

Knowledge Box 2.2

Derivatives of parametric curves

The slope m of the tangent line to a parametric curve $(x(t), y(t))$ at parameter value t is:

$$m = \frac{y'(t)}{x'(t)}$$

Example 2.4   Find the derivative $\dfrac{dy}{dx}$ of the parametric curve with $x(t) = 5\cos(t)$ and $y(t) = 3\sin(t)$.

Solution:

First compute $x'(t) = -5\sin(t)$ and $y'(t) = 3\cos(t)$.

$$\frac{dy}{dx} = \frac{y'(t)}{x'(t)}$$

$$= \frac{-5\sin(t)}{3\cos(t)}$$

$$= -\frac{5}{3}\tan(t)$$

$$\Diamond$$

Example 2.5   Find the tangent line to $(t^2, 2t - 3)$ at (4,1).

Solution:

Our first job is to check if the point is on the curve. If $2t - 3 = 1$, then $t = 2$, and the point for $t = 2$ is (4,1). So, no problem there.

$$\frac{dy}{dx} = \frac{y'(t)}{x'(t)} = \frac{2}{2t} = \frac{1}{t}$$

so we have $m = \dfrac{1}{t} = \dfrac{1}{2}$. Computing the line using the point-slope formula, we get:

$$y - 1 = \frac{1}{2}(x - 4)$$

$$y = \frac{1}{2}x - 2 + 1$$

$$y = \frac{1}{2}x - 1$$

$$\diamondsuit$$

Let's do that again for the practice.

Example 2.6   Find the tangent line to $(4\cos(t), 2\sin(t))$ at $(2\sqrt{3}, 1)$.

Solution:

Our first job is to check if the point is on the curve. If $2\sin(t) = 1$, then $\sin(t) = 1/2$, and so $t = \dfrac{\pi}{6}$. Then $4\cos\left(\dfrac{\pi}{6}\right) = 4\dfrac{\sqrt{3}}{2} = 2\sqrt{3}$. So the point is on the curve, no problem. Compute the slope of the tangent line:

$$\frac{dy}{dx} = \frac{y'(t)}{x'(t)} = \frac{2\cos(t)}{-4\sin(t)} = -\frac{1}{2}\cot(t)$$

At $t = \dfrac{\pi}{6}$ this is $m = -\dfrac{1}{2}\sqrt{3} = -\dfrac{\sqrt{3}}{2}$. Apply the point-slope formula and simplify:

$$y - 1 = -\frac{\sqrt{3}}{2}(x - 2\sqrt{3})$$

$$y = -\frac{\sqrt{3}}{2}x + 3 + 1$$

$$y = -\frac{\sqrt{3}}{2}x + 4$$

and we have the tangent line.

Example 2.7   Find the parameter values for which the tangent line to $(2\sin(t), \cos(t))$ has slope $m = -1/2$.

Solution:

$$\frac{dy}{dx} = \frac{y'(t)}{x'(t)} = \frac{-\sin(t)}{2\cos(t)} = -\frac{1}{2}\tan(t)$$

Solve:

$$-\frac{1}{2}\tan(t) = -\frac{1}{2}$$

$$\tan(t) = 1$$

$$t = \frac{\pi}{4} + n\pi \qquad\qquad n = 0, \pm 1, \pm 2, \ldots$$

Remember that the angles with tangent equal to 1 are the ones with *equal* sine and cosine.

This section has been a small sampler platter of the many sorts of parametric curves that are possible. We've also restricted ourselves to only two coordinates. In Section 2.3 we will revisit this with the formalism of vectors rather than parameters.

PROBLEMS

Problem 2.8   Graph each of the following parametric curves.

1. $(\cos(t), \sin(t) + 2), 0 \leq t \leq 2\pi$

2. $(1 - t, 2t - 4), -5 \leq t \leq 5$

3. $(\sin(t), t\cos(t)), -2\pi \leq y \leq 2\pi$

4. $(t\cos(t), t\sin(t)), 0 \leq t \leq 4\pi$

5. $(2\cos(3t), 3\sin(2t)), 0 \leq t \leq 2\pi$

6. $(t^2, t), -3 \leq t \leq 3$

Problem 2.9   Find a parametric equation for an ellipse with major axis of length 12 in the direction of the y-axis and minor axis of length 2 in the direction of the x-axis.

Problem 2.10   Find a parametric equation for an ellipse with major axis of length 5 in the direction of the x-axis and minor axis of length 3 in the direction of the y-axis.

Problem 2.11   Find a parametric equation for an ellipse with major axis of length 14 in the direction of the y-axis and minor axis of length 7 in the direction of the x-axis.

Problem 2.12   Find the standard $y = mx + b$ form for each of the following parametric lines.

1. $(t, 2t + 1)$

2. $(2t, 1 - t)$

3. $(3t + 1, 5t - 1)$

4. $(t + 7, 7 - t)$

5. $(3t - 1, 3 - 2t)$

6. $(2 - 3t, 3 - 2t)$

Problem 2.13   Graph and carefully describe the parametric curve

$$(2\sin(t), 3\sin(t) + 1)$$

Problem 2.14   Find y' for each of the following parametric curves.

1. $(\sin(2t),\ \cos(3t))$

2. $(1 - t^2,\ 3t + 1)$

3. $(t \cdot \sin(t),\ \cos(t))$

4. $(t \sin(t),\ t \cos(t))$

5. $(\cos(5t),\ 2\sin(3t))$

6. $(t^3,\ t^2)$

Problem 2.15   Find the tangent line to $(t^3,\ t^2)$ at $(-8,\ 4)$.

Problem 2.16   Find the tangent line to $(\sin(t),\ t)$ at $(1,\ \pi/2)$.

Problem 2.17   Find the tangent line to

$$\left(\frac{1}{t^2 + 1},\ \frac{1}{t} \right)$$

at the point $(\frac{1}{2},\ 1)$.

Problem 2.18   For which values of t does

$$(\cos(t),\ \sin(2t))$$

have a tangent line with a slope of 2?

Problem 2.19   Let $f(t) = \dfrac{e^t}{e^t + 1}$. Describe carefully the parametric curve

$$(f(t)\cos(t),\ f(t)\sin(t))$$

for $-\infty < t < \infty$.

Problem 2.20   The parametric curve $(af(t),\ bf(t))$ on $-\infty < t < \infty$ is a line or a line segment. Give explicit directions on how to figure out which line segment it is.

Problem 2.21   Plot carefully the parametric curve:

$$(\sin(t) + \cos(2t)/2,\ \cos(t) + \sin(2t)/2)$$

Problem 2.22   A function cannot assign multiple values to the same point. In light of this, why are points where a parametric curve intersects itself not a problem?

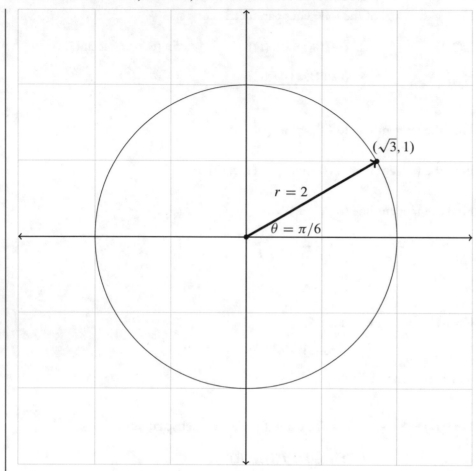

Figure 2.5: Point in both polar and standard coordinates.

2.2 POLAR COORDINATES

Polar coordinates are an alternate coordinate system for doing business, based on direction and distance instead of x-coordinate and y-coordinate. Examine Figure 2.5. The point $(\sqrt{3}, 1)$ can also be specified by going a distance (radius) of two from the origin in the direction of the angle $\theta = \dfrac{\pi}{6}$.

The circle of radius 2 is shown to supply clarity – polar graphs use circles and radial lines the way that a standard graph uses grids. Figure 2.6 shows an example of polar graph paper.

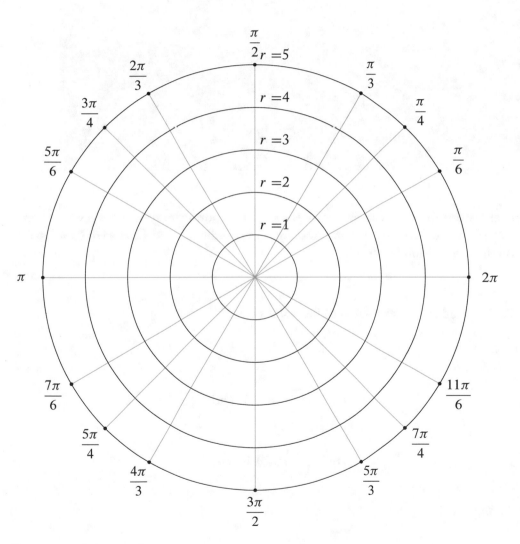

Figure 2.6: Polar graph paper.

Knowledge Box 2.3

Polar to rectangular conversion formulas

In order to transform between (x, y) and (r, θ) coordinates we use the following formulas:

- $x = r \cdot \cos(\theta)$

- $y = r \cdot \sin(\theta)$

- $r = \sqrt{x^2 + y^2}$

- $\theta = \tan^{-1}(y/x)$

Where the formulas don't work, when x is zero and so θ is an odd multiple of $\pi/2$, use common sense – the directions are vertical.

For historical reasons there are two names for the standard coordinate system when we are comparing it to the polar coordinate system – **rectangular** coordinates and **Cartesian** coordinates. We will use the terms interchangeably.

Example 2.23   Find the polar version of the point $(2, 1)$.

Solution:

$$r = \sqrt{2^2 + 1^2} = \sqrt{5}, \text{ and}$$

$$\theta = \tan^{-1}\left(\frac{1}{2}\right) \cong 0.4636 \text{ rad}$$

So the polar point is $(r, \theta) = (\sqrt{5}, 0.4636)$.

◊

Example 2.24   Find the rectangular coordinates for the polar point $\left(4, \dfrac{3\pi}{4}\right)$.

Solution:

In this case $r = 4$ and $\theta = \dfrac{3\pi}{4}$ so:

$$x = r \cdot \cos(\theta) = 4 \cdot \cos\left(\frac{3\pi}{4}\right) = 4 \cdot -\frac{\sqrt{2}}{2} = -2\sqrt{2}$$

$$y = r \cdot \sin(\theta) = 4 \cdot \sin\left(\frac{3\pi}{4}\right) = 4 \cdot \frac{\sqrt{2}}{2} = 2\sqrt{2}$$

and so the corresponding point in Cartesian coordinates is $(x, y) = (-2\sqrt{2}, 2\sqrt{2})$.

◊

One of the major uses for converting between the two coordinate systems is to permit us to plot polar points on normal graph paper when we are graphing a polar function. If you have a good eye, or a protractor, it is possible to plot polar points directly, but typically a person just learning polar coordiantes have far more practice plotting (x, y)-points.

One of the nice things about polar coordinates is that they let us deal very easily with circles centered at the origin. Circles centered at the origin are *constant functions* in polar coordinates. The next example demonstrates this.

Example 2.25   Graph the polar function $r = 2$.

Solution:

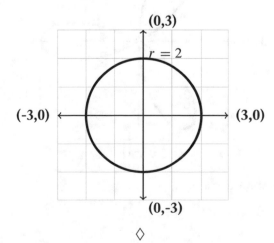

◊

We usually write polar functions in the form

$$r = f(\theta)$$

making the angle the independent variable and the r the dependent variable. This makes it very easy to give polar functions as parametric functions.

Knowledge Box 2.4

Parametric form of polar curves

If $r = f(\theta)$ on $\theta_1 \leq \theta \leq \theta_2$ is a polar curve, then a parametric form for the same curve is:

$$(f(t) \cdot \cos(t), \ f(t) \cdot \sin(t))$$

for $t \in [\theta_1, \theta_2]$.

So far in this section we have established the connections between polar coordinates and the rest of the systems developed in this text. It is time to display polar curves that have unique characteristics that are most easily seen in the polar system.

Example 2.26   Graph the polar function $r = \cos(3\theta)$ on $[0, \pi)$.

Solution:

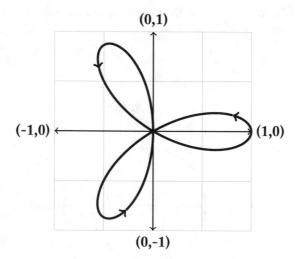

The arrows show the drawing direction.

◊

Definition 2.3 Petal curves *are curves with equations of the form:*

$$r = \cos(n\theta) \ or \ r = \sin(n\theta),$$

where n is an integer.

If no restriction is placed on n, then the curve is traced out an infinite number of times. For odd n, a domain of $\theta \in [0, \pi)$ traces the curve once; when n is even, $\theta \in [0, 2\pi)$ is needed to trace the entire curve once.

Example 2.27   Compare the curves $r = \sin(5\theta)$ and $r = \cos(5\theta)$ on the range $\theta \in [0, \pi)$.

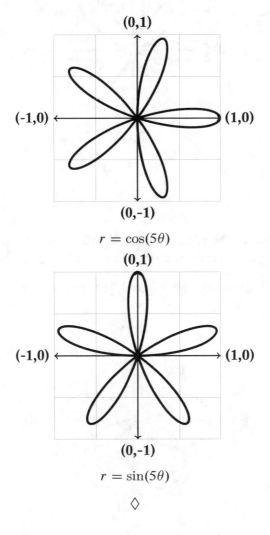

$r = \cos(5\theta)$

$r = \sin(5\theta)$

◊

The odd fact, that the minimal domain (to hit all the points) is twice as large when n is even, is to some degree explained by the fact that, while petal curves with odd parameter n yield n petals, when n is even we get $2n$ petals.

Example 2.28   Plot the polar function $r = \cos(4\theta)$.
Solution:

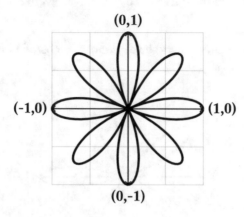

See? We have $n = 4$ but 8 petals.

◇

It is possible to use values of the petal-determining parameter for polar curves that are not integers, but then figuring out the minimal domain to plot the curve becomes problematic.

Example 2.29   Plot the polar function $r = \cos(1.5\,\theta)$ for $\theta \in [0, 4\pi)$.

Solution:

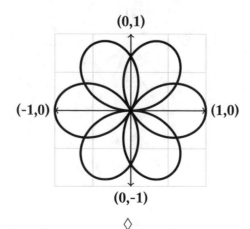

◇

A fractional parameter yields more and fatter petals on the curve. If you have polar curve plotting software, experiment with these parameters and see what you can find.

2.2.1   POLAR CALCULUS

There are many interesting shapes that can be made with polar coordinates, but at this point we are going to look into the calculus of polar curves. First, computing $\dfrac{dy}{dx}$ at a point (r, θ).

We already have a parametric form for polar curves:

$$x(\theta) = f(\theta) \cdot \cos(\theta), \quad y(\theta) = f(\theta) \cdot \sin(\theta)$$

This is the parametric curve:

$$(x = f(\theta) \cos(\theta), \; y = f(\theta) \sin(\theta))$$

We can then apply the technique for finding the derivative of a parametric curve and get:

$$\frac{dy}{dx} = \frac{y'(\theta)}{x'(\theta)} = \frac{f'(\theta) \cdot \sin(\theta) + f(\theta) \cdot \cos(\theta)}{f'(\theta) \cdot \cos(\theta) - f(\theta) \cdot \sin(\theta)}$$

That's not a compact or elegant formula, but it is a formula.

Let's put this formula in a Knowledge Box.

Knowledge Box 2.5

Computing $\dfrac{dy}{dx}$ for a polar curve

$$\frac{dy}{dx} = \frac{f'(\theta) \cdot \sin(\theta) + f(\theta) \cdot \cos(\theta)}{f'(\theta) \cdot \cos(\theta) - f(\theta) \cdot \sin(\theta)}$$

Example 2.30   Find the tangent line to $r = \cos(3\theta)$ at $\theta = \dfrac{\pi}{12}$.

Solution:

To use the formula in Knowledge Box 2.5 we need $f(\theta) = r = \cos(3\theta)$ and $f'(\theta) = -3\sin(3\theta)$. Computing the derivative we get:

$$\frac{dy}{dx} = \frac{-3\sin(3\theta)\sin(\theta) + \cos(3\theta)\cos(\theta)}{-3\sin(3\theta)\cos(\theta) - \cos(3\theta)\sin(\theta)}$$

$$= \frac{-3\sin\left(3\frac{\pi}{12}\right)\sin\left(\frac{\pi}{12}\right) + \cos\left(3\frac{\pi}{12}\right)\cos\left(\frac{\pi}{12}\right)}{-3\sin\left(3\frac{\pi}{12}\right)\cos\left(\frac{\pi}{12}\right) - \cos\left(3\frac{\pi}{12}\right)\sin\left(\frac{\pi}{12}\right)} \cong -0.06$$

When $\theta = \dfrac{\pi}{12}$ we get the point in polar coordinates $\left(\cos\left(\dfrac{\pi}{4}\right), \dfrac{\pi}{12}\right) = \left(\dfrac{\sqrt{2}}{2}, \dfrac{\pi}{12}\right)$. Now we need the Cartesian version of the point. Applying the polar-to-rectangular formulas we obtain:

$(\dfrac{\sqrt{2}}{2}\cos\left(\dfrac{\pi}{12}\right), \dfrac{\sqrt{2}}{2}\sin\left(\dfrac{\pi}{12}\right)) \cong (0.683, 0.183)$

Find the line:

$$y - 0.183 = -0.06(x - 0.683)$$
$$y = -0.06x + 0.224$$

Let's actually take a look at the graph with the curve and the tangent line. The tangent line hits the curve at three points, but the point of tangency is also plotted and is clearly a tangent line; the other two intersections cut through the curve at sharp angles.

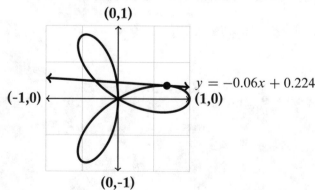

The curve $r = \cos(3\theta)$ and the tangent line at $\theta = \dfrac{\pi}{12}$.

◇

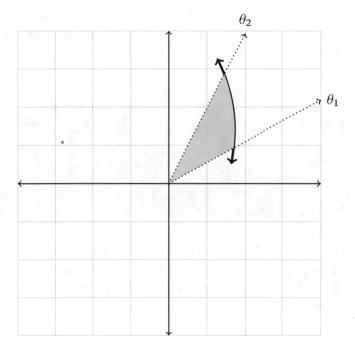

Figure 2.7: Area "under the curve" in polar coordinates.

The formula for the Cartesian derivative of a curve in polar coordinates is pretty cumbersome, but the Cartesian form of $r = \cos(3\theta)$ is really just not manageable at all. So, if we need tangent lines to these curves, this formula is our only hope.

What about integrals? If we have a sector of a circle subtending an angle of θ, then the area is just the fraction of the full circle that θ covers times the area formula of πr^2 so we get

$$\text{Area} = \frac{\theta}{2\pi} \cdot \pi r^2 = \frac{\theta}{2} r^2.$$

This means that the farther from the origin an area is, the larger it gets. This leads to a really different integral formula for area. First of all, it is area enclosed by the polar curve. Second, we need to be very careful about $r < 0$ because it's so much less obvious than $y < 0$.

The shaded area between the polar curve shown in Figure 2.7 and the origin, in the angular range $\theta_1 \leq \theta \leq \theta_2$ is

$$\text{Area} = \frac{1}{2} \int_{\theta_1}^{\theta_2} r(\theta)^2 \cdot d\theta$$

When we did integrals in Cartesian space, we approximated areas with rectangles. In polar space, we use pie-shaped slices, and so the area depends on the square of the functional value instead of just its value. The constant of one-half carries over from the area formulas for sectors of a circle.

Knowledge Box 2.6

Finding the area between a polar curve and the origin

$$Area = \frac{1}{2} \int_{\theta_1}^{\theta_2} r(\theta)^2 \cdot d\theta$$

In Chapter 1, we had to be careful about area above and below the x-axis. When working with polar curves, the analogous problems are areas with $r \geq 0$ and $r \leq 0$. If we know r stays positive, then we still need to know the minimal angular domain that sweeps out the entire shape once. Let's do some examples.

Example 2.31   Find the area enclosed by $r = \theta$ from $\theta = 0$ to $\theta = \dfrac{4\pi}{3}$.

Solution: r is positive on the domain of integration so we can just integrate.

$$A = \frac{1}{2} \int_0^{4\pi/3} r^2 \cdot d\theta$$

$$= \frac{1}{2} \int_0^{4\pi/3} \theta^2 \cdot d\theta$$

$$= \frac{1}{6} \theta^3 \Big|_0^{4\pi/3}$$

$$= \frac{1}{6} \left(\left(\frac{4\pi}{3} \right)^3 - 0 \right)$$

$$= \frac{32\pi^3}{81} \text{ units}^2$$

$$\cong 12.25 \text{ units}^2$$

Let's look at a picture of the area:

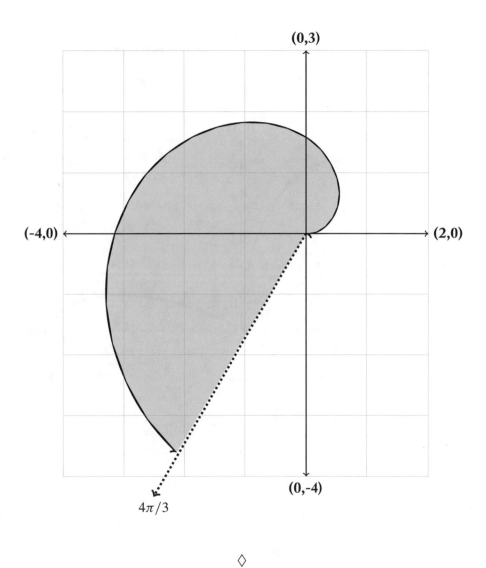

$\Diamond$

Now for a more challenging example. The petal curve function from Example 2.26 was $r = \cos(3\theta)$ on $[0, \pi)$. It has both positive and negative r in the course of the graph.

Example 2.32   Find the area enclosed by $r = \cos(3\theta)$.

Solution:

Let's graph this again, using different shades for the positive and negative r.

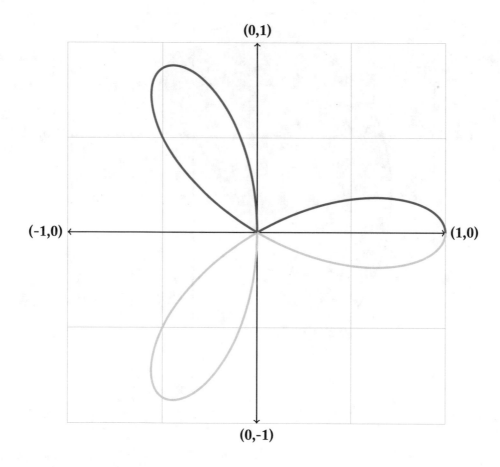

Notice the first dark gray arc of the graph is from $\theta = 0$ to $\theta = \pi/6$. This arc encompasses half of one of the three petals and so is $\dfrac{1}{6}$ of the area. Since on this angular interval r is positive, computing six times the integral of $r = \cos(3\theta)$ on $[0, \pi/6]$ will give us the full area of the curve.

$$\text{Area} = 6 \cdot \frac{1}{2} \int_0^{\pi/6} \cos^2(3\theta) \cdot d\theta$$

$$= 3 \int_0^{\pi/6} \frac{1 + \cos(6\theta)}{2} \cdot d\theta$$

$$= \frac{3}{2} \int_0^{\pi/6} d\theta + \frac{3}{2} \int_0^{\pi/6} \cos(6\theta) \cdot d\theta$$

$$= \frac{3}{2} \int_0^{\pi/6} d\theta + \frac{3}{2} \int_0^{\pi} \cos(\theta) \cdot d\theta \qquad \text{Cosine was rescaled}$$

$$= \frac{3}{2} \theta \Big|_0^{\pi/6} + \frac{3}{2} \sin(\theta) \Big|_0^{\pi}$$

$$= \frac{3}{2} \left(\frac{\pi}{6} - 0 \right) + \frac{3}{2}(0 - 0)$$

$$= \frac{\pi}{4} \text{ units}^2$$

PROBLEMS

Problem 2.33   For each of the following points in Cartesian coordinates, give the corresponding points in polar coordinates.

1. (1,1)

2. (3,4)

3. $(4, 4\sqrt{3})$

4. (-3,7)

5. (-5,-5)

6. (0,8)

Problem 2.34   For each of the following points in polar coordinates, give the corresponding points in rectangular coordinates. The points are in the form (r, θ) with the angular coordinate in radians.

1. $(4, \pi/3)$ 3. $(-5, \pi/4)$ 5. $(0.2, 3\pi/4)$

2. $(2, \pi/5)$ 4. $(7, 4\pi/3)$ 6. $(2.6, 1.14)$

Problem 2.35   Plot the following polar curves on the given domain.

1. $r = \cos(2\theta)$ on $[0, 2\pi)$ 4. $r = 4(\cos(\theta) + 1)$ on $[0, 2\pi)$

2. $r = \sin(7\theta)$ on $[0, \pi)$ 5. $r = \theta/4$ on $[0, 6\pi)$

3. $r = \sin(3\theta) + 2$ on $[0, 2\pi)$ 6. $r = 5\cos(\theta)$ on $[0, \pi)$

Problem 2.36   Find a parametric form for $3 = 5\cos(\theta)$ on $[0, \pi)$.

Problem 2.37   Find the tangent line to $r = 6$ at $\theta = \dfrac{3\pi}{4}$.

Problem 2.38   Find the tangent line to $r = \theta$ at $\theta = \dfrac{\pi}{4}$.

Problem 2.39   Find the tangent line to $r = \cos(4\theta)$ at $\theta = \dfrac{\pi}{16}$.

Problem 2.40   Using the polar integral, verify that a circle of radius R has area equal to πR^2.

Problem 2.41   Graph $r = \sin(5\theta)$ on $[0, \pi)$ using two colors or line styles to show where r is positive or negative.

Problem 2.42   Find the area enclosed by $r = \sin(5\theta)$. Worry about the issue of negative radii.

Problem 2.43   Graph $r = \cos(2\theta)$ on $[0, 2\pi)$ using two colors or line styles to show where r is positive or negative.

Problem 2.44   Find the area enclosed by $r = \cos(2\theta)$. Worry about the issue of negative radii.

Problem 2.45   Using software of some sort, graph $r = \cos(1.1 \cdot \theta)$ on $[0, 20\pi)$.

Problem 2.46   Using software of some sort, graph $r = \sin(5.1 \cdot \theta)$ on $[0, 20\pi)$.

Problem 2.47   Describe completely all polar curves of the form $r = C \cdot \sin(\theta)$ in simple English.

Problem 2.48   For the polar curve $r = \theta$ find the angle α so that the area between the curve and the origin on the interval $[0, \alpha]$ is 2.

Problem 2.49   Carefully graph the polar curve $r = \cos^2(\theta)$.

Problem 2.50   Carefully graph the polar curve $r = \sin^2(\theta)$.

Problem 2.51   Why is the curve

$$r = 3 - 3\sin(\theta)$$

called a *cartiod*?

Problem 2.52   Write a paragraph accurately explaining where the term "Cartesian," as used in describing a coordinate system, comes from.

Problem 2.53   For the polar curve $r = \cos A\theta$ where A is a decimal number with a positive, finite number of non-zero decimals, find the rule for the domain for θ that permits the curve to be graphed once.

2.3   VECTOR FUNCTIONS

One of the most unsatisfying things about the treatment of position, velocity, and acceleration in single-variable calculus was that these things happened in one dimension. This situation can be amended by using **vectors**. A vector is a lot like a point – it has coordinates and a dimension – but a vector is thought of as specifying a direction and a magnitude rather than just a point in space.

Example 2.54   A vector:

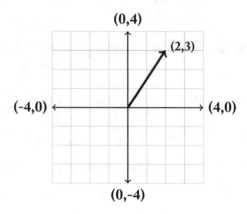

The vector $\vec{v} = (2, 3)$

We use a small arrow over a variable name to denote that an object is a vector. So, the vector shown above is $\vec{v} = (2, 3)$. In a way, vectors are like multi-dimensional numbers that let us work in as many dimensions as we need to. Vectors have their own arithmetic. It's called vector arithmetic.

Knowledge Box 2.7

Vector Arithmetic

If $\vec{v} = (v_1, v_2, \ldots, v_n)$ and $\vec{w} = (w_1, w_2, \ldots, w_n)$ are vectors in n–dimensions and c is a constant, then we define the following.

- $c \cdot \vec{v} = (cv_1, cv_2, \ldots, cv_n)$ *(scalar multiplication)*

- $\vec{v} + \vec{w} = (v_1 + w_1, v_2 + w_2, \ldots, v_n + w_n)$ *(vector addition)*

- $\vec{v} - \vec{w} = (v_1 - w_1, v_2 - w_2, \ldots, v_n - w_n)$ *(vector subtraction)*

- $\vec{v} \cdot \vec{w} = v_1 w_1 + v_2 w_2 + \ldots + v_n w_n$ *(dot product)*

Example 2.55   If $\vec{v} = (1, 2, -1)$ and $\vec{w} = (0, 2, 4)$, compute $5 \cdot \vec{w} - \vec{v}$.

Solution:

$$5 \cdot (0, 2, 4) - (1, 2, -1) = (0, 10, 20) - (1, 2, -1) = (0 - 1, 10 - 2, 20 - (-1)) = (-1, 8, 21)$$

$$\Diamond$$

Example 2.56   If $\vec{v} = (-2, 1, 3)$ and $\vec{w} = (1, 2, 1)$, compute $\vec{v} \cdot \vec{w}$.

Solution:

$$\vec{v} \cdot \vec{w} = -2 \cdot 1 + 1 \cdot 2 + 3 \cdot 1 = -2 + 2 + 3 = 3$$

$$\Diamond$$

There are a number of useful algebraic rules for vector arithmetic.

Knowledge Box 2.8

Vector Algebra

- $c \cdot (\vec{v} + \vec{w}) = c \cdot \vec{v} + c \cdot \vec{w}$
- $c \cdot (d \cdot \vec{v}) = (cd) \cdot \vec{v}$
- $\vec{v} + \vec{w} = \vec{w} + \vec{v}$
- $\vec{u} \cdot (\vec{v} + \vec{w}) = \vec{u} \cdot \vec{v} + \vec{u} \cdot \vec{w}$

Definition 2.4 *The* **length** *or* **magnitude** *of a vector* $\vec{v} = (v_1, v_2, \ldots, v_n)$, *denoted* $|\vec{v}|$, *is given by*

$$|\vec{v}| = \sqrt{v_1^2 + v_2^2 + \cdots + v_n^2}$$

Notice that this is similar to the formula for calculating distance. The length of a vector is the distance it spans, so this is a natural definition.

It turns out that there is a really useful property of one way we multiply two vectors, the dot product in Knowledge Box 2.7, that also uses the length of vectors. Suppose we have two vectors $\vec{v}$ and $\vec{w}$ as shown in Figure 2.8. Then, we can use the dot product together with the lengths of the vectors to calculate the angle between them.

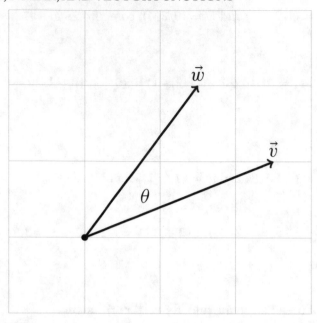

Figure 2.8: Two vectors.

Knowledge Box 2.9

Formula for the angle between vectors

$$\cos(\theta) = \frac{\vec{v} \cdot \vec{w}}{|v||w|}$$

What makes this property so useful? If two vectors are at right angles, then the cosine of the angle between them is zero. This means we can detect right angles in a new and easy way and in any number of dimensions. In particular, **two vectors are at right angles to one another if and only if their dot product is zero**.

Example 2.57   Which of the following pairs of vectors are at right angles to one another?

1. $\vec{v} = (1, 1)$ and $\vec{w} = (1, -1)$

2. $\vec{s} = (1, 2, 1)$ and $\vec{t} = (-1, 2, -3)$

3. $\vec{a} = (3, 1, 2)$ and $\vec{b} = (2, -2, 3)$

Solution:

Compute dot products:

$$\vec{v} \cdot \vec{w} = (1, 1) \cdot (1, -1) = 1 - 1 = 0$$

So $\vec{v}$ and $\vec{w}$ are at right angles to one another.

$$\vec{s} \cdot \vec{t} = (1, 2, 1) \cdot (-1, 2, -3) = -1 + 4 - 3 = 0$$

So $\vec{s}$ and $\vec{r}$ are at right angles to one another.

$$\vec{a} \cdot \vec{b} = (3, 1, 2) \cdot (2, -2, 3) = 6 - 2 + 6 = 10$$

So $\vec{a}$ and $\vec{b}$ are not at right angles to one another.

Definition 2.5 *Objects at right angles to one another are also said to be* **orthogonal** *to one another.*

So with our new word we can re-state this property of the dot product as follows. **Two vectors are orthogonal if and only if they have a dot product of zero.**

Definition 2.6 *The* **cross product** *of* $\vec{v} = (v_1, v_2, v_3)$ *and* $\vec{w} = (w_1, w_2, w_3)$ *is defined to be:*

$$\vec{v} \times \vec{w} = (v_2 w_3 - v_3 v_2, \quad v_3 w_1 - v_1 w_3, \quad v_1 w_2 - v_2 w_1)$$

The cross product of two vectors is only defined in three dimensions. It is a very special purpose operator, but it is quite useful in physics. The following property is used to construct systems of directions.

Knowledge Box 2.10

Orthogonality of the cross product

If $\vec{v}$ and $\vec{w}$ are non-zero vectors in three dimensions that are not scalar multiples of one another, then $\vec{v} \times \vec{w}$ is at right angles to (orthogonal to) both $\vec{v}$ and $\vec{w}$.

Example 2.58   Find a vector at right angles to both (1, 1, 2) and (3, 0, 4). Check your answer by taking the relevant dot products.

Solution:

Use the cross product:

$$(1,1,2) \times (3,0,4) = (1 \cdot 4 - 2 \cdot 0, 2 \cdot 3 - 1 \cdot 4, 1 \cdot 0 - 1 \cdot 3) = (4,2,-3)$$

Check:

$$(1,1,2) \cdot (4,2,-3) = 4 + 2 - 6 = 0 \checkmark$$

$$(3,0,4) \cdot (4,2,-3) = 12 + 0 - 12 = 0 \checkmark$$

So the vector $(4,2,-3)$ *is* orthogonal to $(1,1,2)$ and $(3,0,4)$.

$$\Diamond$$

Definition 2.7 *A* **unit vector** *is a vector $\vec{v}$ with $|\vec{v}| = 1$.*

This is an almost trivial definition, but, coupled with the next Knowledge Box rule, it captures an important feature of vectors.

Knowledge Box 2.11

The unit vector in the direction of a given vector

If $\vec{v}$ is not zero, then

$$\frac{1}{|\vec{v}|} \cdot \vec{v}$$

is the unit vector in the direction of $\vec{v}$.

Unit vectors have a number of applications, but the one we will use the most often is that they capture a notion of **direction**. There are an infinite number of vectors in any given direction, but only two unit vectors, and they point in opposite directions from one another.

Example 2.59   Find the unit vector in the direction of $\vec{v} = (1, 2)$.

Solution:

Start with $|\vec{v}| = \sqrt{1^2 + 2^2} = \sqrt{5}$. Then the desired unit vector is

$$\frac{1}{\sqrt{5}} \cdot (1, 2) = \left(\frac{1}{\sqrt{5}}, \frac{2}{\sqrt{5}}\right)$$

◊

We've now reviewed enough vector algebra to start working with vector functions.

Definition 2.8 *A* **vector function** *is a vector $\vec{v}(t)$ whose coordinates are functions of a parameter t.*

This is very close to the notion of a parametric curve. In fact, for some applications the two notions are interchangeable.

Example 2.60   Identify the points that appear on the graph of the vector function

$$\vec{v}(t) = (\cos(t), \sin(t))$$

Solution:

The unit circle centered at the origin.

◊

Vector functions permit us to specify a large number of different types of curves through space. They share with parametric curves the property that they can encode systems that are not Cartesian functions at all.

Example 2.61   Plot the vector curve $\vec{w}(t) = (\sin(t) + t/5, \cos(t) + t/8)$.

Solution:

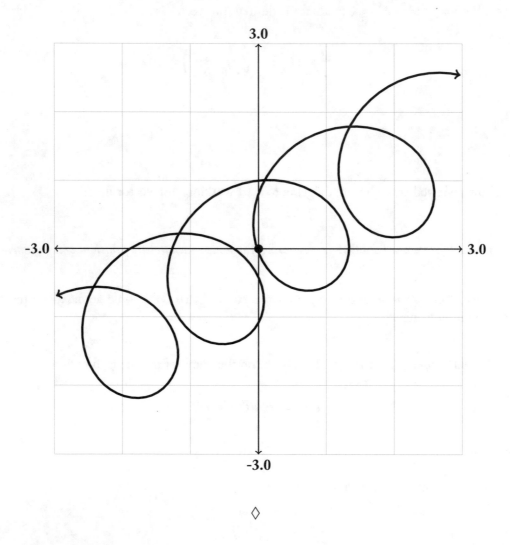

$\Diamond$

It is not too hard to understand this curve. The sine and cosine terms make a simple circle while the $t/5$ and $t/8$ terms create a line that moves the center of the circle. We can play a similar trick by combining two circles. We will make the circles both have distinct radii and have the particle travel around them at different speeds.

Example 2.62   Plot the vector curve $\vec{w}(t) = (2\sin(t/4) + \cos(t), 2\cos(t/4) + \sin(t))$.

Solution:

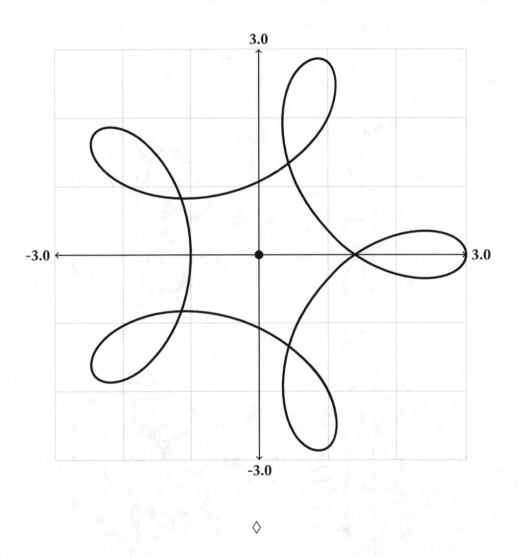

This curve is tracing a circle of radius 2 that moves at one fourth the rate of a circle of radius one that is circling the current point of the circle of radius 2. It is potentially instructive to look at the previous example for several different speeds of motion of the particle on the smaller circle.

Example 2.63   Plot the vector curves:

- $\vec{w}(t) = (2\sin(t/4) + \cos(t), 2\cos(t/4) + \sin(t))$

- $\vec{v}(t) = (2\sin(t/4) + \cos(2t), 2\cos(t/4) + \sin(2t))$

- $\vec{s}(t) = (2\sin(t/4) + \cos(3t), 2\cos(t/4) + \sin(3t))$

- $\vec{t}(t) = (2\sin(t/4) + \cos(4t), 2\cos(t/4) + \sin(4t))$

Solution:

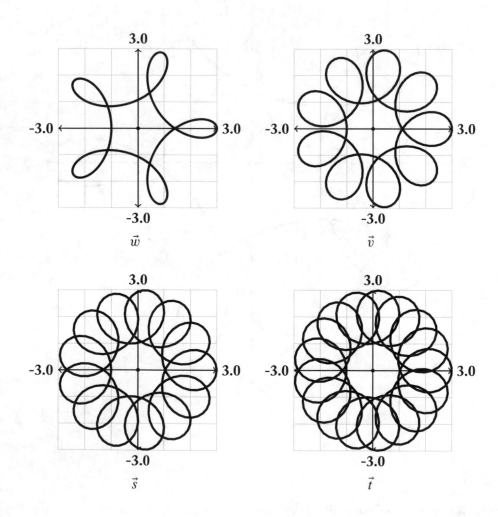

$\Diamond$

2.3.1   CALCULUS WITH VECTOR CURVES

A big advantage of vector functions is that they free us from the tyranny of abstract one-dimensional position, velocity, and acceleration.

Definition 2.9 *If* $\vec{v}(t) = (f_1(t), f_2(t), \ldots, f_n(t))$ *then the* **derivative** *of* $\vec{v}(t)$ *is*

$$\vec{v}'(t) = (f_1'(t), f_2'(t), \ldots, f_n'(t)).$$

Here is the payoff. If $\vec{s}(t)$ gives the position of a particle, then its velocity vector is $\vec{v}(t) = \vec{s}'(t)$, and its acceleration vector is $\vec{a}(t) = \vec{v}'(t)$. The derivative-based relationships between position, velocity, and acceleration hold for vector functions just as they do for ordinary functions. The difference is that we may now describe motion in complex two- and three-dimensional paths.

Example 2.64   Suppose the path of a particle is given by the vector function:

$$\vec{s}(t) = (\sin(t), \cos(2t))$$

Plot the particle's path and find its velocity and acceleration vectors.

Solution:

The plot looks like this:

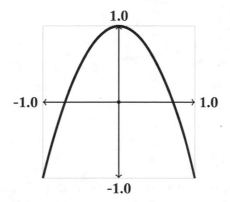

Take derivatives to find

$$\vec{v}(t) = (\cos(t), -2\sin(2t)) \text{ and } \vec{a}(t) = (-\sin(t), -4\cos(2t))$$

◊

What do the vector velocity and acceleration mean? They give (as their magnitude) the *amount* of velocity or acceleration, but they also tell us which direction in space the velocity or acceleration is going.

Example 2.65   If a particle has position $\vec{s}(t) = (2\cos(t), \sin(t))$, plot the position curve and show the velocity vectors – starting at the curve – at times $t = \pi/6, 2\pi/3, 7\pi/6$, and $5\pi/3$.

Solution:

Taking a derivative we get that the velocity vector is $\vec{v} = (-2\sin(t), \cos(t))$. So, let's plot the curve and the vectors at the specified times.

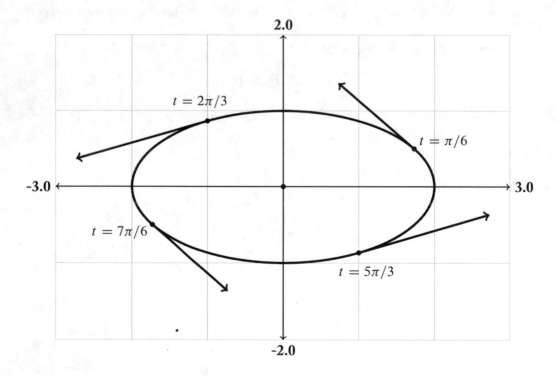

Notice that these vectors are tangent to the curve and also show which direction the curve is oriented. They show the instantaneous velocity of the particle following the curve.

◊

Example 2.66   Find the times when the position curve $\vec{s}(t) = (2\sin(t), \cos(t) + 1)$ has velocity of the greatest magnitude.

Solution:

First, take a derivative to obtain the velocity function:

$$\vec{v}(t) = (2\cos(t), -\sin(t))$$

The magnitude of the velocity is then

$$|(2\cos(t), -\sin(t))| = \sqrt{4\cos^2(t) + \sin^2(t)} = \sqrt{3\cos^2(t) + 1}$$

Note that the second step above used the Pythagorean identity, $\sin^2(t) + \cos^2(t) = 1$. Now we need to know when this is largest... Since $0 \le \cos^2(t) \le 1$ it will simply happen when $\cos^2(t) = 1$. This means that $\cos(t) = \pm 1$. So the times are $t = k\pi$ for any whole number k.

Note that the optimization of the function

$$velocity = \sqrt{3\cos^2(t) + 1}$$

was *not* performed by solving a derivative equal to zero. Rather, reasoning based on well know propeties of the function cosine were used to extract the maximum. Again: mathematics is the art of *avoiding* calculation.

PROBLEMS

Problem 2.67   For each possible pair of the following four vectors, say which are orthogonal.

- $\vec{v} = (1, 1)$
- $\vec{r} = (1, 2)$
- $\vec{s} = (2, -1)$
- $\vec{w} = (5, -5)$

Problem 2.68   For each possible pair of the following four vectors, say which are orthogonal.

- $\vec{v} = (1, 1, 1)$
- $\vec{r} = (-2, 2, 0)$
- $\vec{s} = (3, 1, -4)$
- $\vec{w} = (1, -2, 1)$

Problem 2.69   Plot the following vector curves. You are free to use software.

1. $\vec{v}(t) = (3\sin(t/3) + \cos(t), \ 3\cos(t/2) + \sin(t))$

2. $\vec{w}(t) = (3\sin(t/3) + \cos(2t), \ 3\cos(t/2) + \sin(2t))$

3. $\vec{s}(t) = (2\sin(t/3) + 2\cos(2t), \ 2\cos(t/2) + 2\sin(2t))$

4. $\vec{q}(t) = (\sin(t/2) + 3\cos(3t), \ \cos(t/2) + 3\sin(3t))$

5. $\vec{a}(t) = (3\cos(2t), \ 3\sin(t))$

6. $\vec{b}(t) = (t + \sin(t), \ \cos(t))$

Problem 2.70   Example 2.64 shows the graph of the position function

$$\vec{s}(t) = (\sin(t), \ \cos(2t))$$

and it looks like a part of a parabola. Demonstrate that it is a parabola by finding the Cartesian form, including the domain on which the curve is defined.

Problem 2.71   If $\vec{s}(t) = (2t, \ \cos(t))$ is the position of a point, then plot the curve traced by the point on $[0, 2\pi]$ and show the velocity vectors at each multiple of $\pi/6$ in the interval.

Problem 2.72   For each of the following position vectors, find the velocity and acceleration vectors.

1. $\vec{s}(t) = (3t + 5, \ 2t + 6)$

2. $\vec{s}(t) = (t^2 + t + 1, \ 5 - t)$

3. $\vec{s}(t) = (\sin(t), \ \cos(2t))$

4. $\vec{s}(t) = (\tan^{-1}(t), \ t^2)$

5. $\vec{s}(t) = (1 + t + t^2, \ 1 - t + t^2)$

6. $\vec{s}(t) = (t\cos(t), \ t\sin(t))$

Problem 2.73   Find when a particle whose position is given by

$$\vec{s}(t) = (\sin(t + \pi/3), \ \cos(t))$$

is traveling parallel to the y-axis.

Problem 2.74   Find a Cartesian function with the same graph as:

$$\vec{w}(t) = (3t + 1, \ 5t + 2)$$

Problem 2.75 Find a Cartesian function with the same graph as

$$\vec{u}(t) = (t^2, \, 1 - 3t)$$

or give a reason it is impossible.

Problem 2.76 Show, by argumentation, that a vector function, all of whose components are linear functions of t, has the same graph as a Cartesian line.

Problem 2.77 The vector function $(\cos(t), \, \sin(t), t)$ is a twisting path. First explain what curve this vector function traces out if we ignore the third coordinate and then do your best to explain what the shape of the curve is.

Problem 2.78 If $f(t)$ is a function with domain $-\infty < t < \infty$, then describe the vector function $(f(t), \, f(t), \, f(t))$ as best you can.

CHAPTER 3

The Arithmetic, Geometry, and Calculus of Polynomials

Polynomials are a rich family of functions. They include lines and quadratic equations – functions that we've studied in a good deal of detail because they are useful for so many things. Polynomials are defined everywhere on the real line; they are continuous, differentiable, and just generally nice to work with. As we will see in our later study of sequences and series, polynomials can be used to approximate any continuous function. In this chapter we will learn to work with polynomials more closely as well as introducing a couple of techniques that are generally useful in the context of polynomials.

In Section 3.1 we present *Newton's method* which is a general technique for finding the roots of an equation. In Section 3.3 we use polynomials to demonstrate a rule, La'Hospital's rule, that is useful for evaluating limits in general. This section should be covered even if the other material on polynomials is not of interest.

3.1 POLYNOMIAL ARITHMETIC

Consider the problem of computing $(x^2 + x + 2)^2$. We might do this in the following fashion:

$$(x^2 + x + 2)(x^2 + x + 2) = x^2(x^2 + x + 2) + x(x^2 + x + 2) + 2(x^2 + x + 2)$$

$$= x^4 + x^3 + 2x^2 + x^3 + x^2 + 2x + 2x^2 + 2x + 4$$

$$= x^4 + 2x^3 + 5x^2 + 4x + 4$$

Which is a lot of algebra! If we made the polynomial higher degree it would get worse. Let's try this instead.

$$
\begin{array}{cccccc}
 & & 1 & 1 & 2 & & (1x^2 + 1x + 2) \\
 & & 1 & 1 & 2 & & (1x^2 + 1x + 2) \\
\hline
 & & 2 & 2 & 4 & & \\
 & 1 & 1 & 2 & \cdot & & \\
1 & 1 & 2 & \cdot & \cdot & & \\
\hline
1 & 2 & 5 & 4 & 4 & & (1x^4 + 2x^3 + 5x^2 + 4x4)
\end{array}
$$

Which also shows $(x^2 + x + 2)^2 = x^4 + 2x^3 + 5x^2 + 4x + 4$. But how? In the "long multi-plication" table the positions correspond to powers of x. The rightmost place is the 1's place, the next is the x's place, then the x^2's place and so on.

It is very similar to the algorithm for multiplying numbers with pencil-and-paper, but *there is no carry*. Why no carry? Because when you add up a whole lot of terms of the form ax^m you just get many x^m's; it never piles up to the point where you get some x^{m+1}'s. In other words, this is actually easier than the multiplication algorithm for numbers.

Knowledge Box 3.1

Fast Polynomial Multiplication

- *Place the coefficients of the polynomials in the first two rows.*

- *In the following rows, scale the coefficients of the first polynomial by each of the coefficients of the second, one per row.*

- *The first new row lines up with the ones above it; after that shift left one place for each row.*

- *Add up the columns for the rows you just generated.*

- *This creates a last row that is the coefficients of the product.*

Example 3.1   Find
$$(x^3 + 4x^2 + 2x - 3) \times (x^2 + 3x + 2)$$

Solution:

$$
\begin{array}{rrrrrr}
 & & 1 & 4 & 2 & -3 \\
 & & & 1 & 3 & 2 \\
\hline
 & & 2 & 8 & 4 & -6 \\
 & 3 & 12 & 6 & -9 & . \\
1 & 4 & 2 & -3 & . & . \\
\hline
1 & 7 & 16 & 11 & -5 & -6 \\
\end{array}
$$

So:
$$(x^3 + 4x^2 + 2x - 3) \times (x^2 + 3x + 2) = x^5 + 7x^4 + 16x^3 + 11x^2 - 5x - 6$$

◇

This technique is enormously faster than imposing the distributive law and collecting terms, and because it's structured, it is easier to avoid errors. There is one potential sand trap – when a term is missing, you must fill in a zero for it. Let's do another example that demonstrates this.

Example 3.2   Find $(x^2 + x + 1) \times (x^2 + 4)$.

Solution:

$$
\begin{array}{rrrrr}
  &   & 1 & 1 & 1 \\
  &   & 1 & 0 & 4 \\
\hline
  &   & 4 & 4 & 4 \\
  & 0 & 0 & 0 & . \\
1 & 1 & 1 & . & . \\
\hline
1 & 1 & 5 & 4 & 4 \\
\end{array}
$$

So:

$$(x^2 + x + 1) \times (x^2 + 4) = x^4 + x^3 + 5x^2 + 4x + 4$$

$$\Diamond$$

A similar technique – that is more likely to be familiar to you – can also be used to divide polynomials. It is sometimes called **synthetic division**.

Example 3.3   Compute $(x^3 + 6x^2 + 11x + 6) \div (x + 2)$.

Solution:

$$
\begin{array}{rr|rrr}
  &   & 1 & 4 & 3 \\
\cline{3-5}
1 & 2 & 1 & 6 & 11 & 6 \\
  &   & 1 & 2 \\
\cline{3-4}
  &   &   & 4 & 11 & 6 \\
  &   &   & 4 & 8 \\
\cline{4-5}
  &   &   &   & 3 & 6 \\
  &   &   &   & 3 & 6 \\
\cline{5-6}
  &   &   &   &   & 0 \\
\end{array}
$$

So:

$$(x^3 + 6x^2 + 11x + 6) \div (x + 2) = x^2 + 4x + 3$$

$$\Diamond$$

These techniques for multiplication and division of polynomials are essentially book-keeping devices. They don't give you new capabilities, but they make existing capabilities more reliable.

Remember that we often want to find the roots of a function. For instance, when optimizing or sketching curves, we want to to be able to solve $f(x) = 0$, $f'(x) = 0$, and $f''(x) = 0$. Being able to rapidly multiply and divide polynomials makes these tasks easier. The **root-factor theorem** tells us that "If $f(x)$ is a polynomial and $f(c) = 0$ for some number c, then $(x - c)$ is a factor of $f(x)$." One thing these new techniques do not do is tell us what value of c to try. Plugging in values of c and looking for zeros often works – if there is some whole number c that is a root. Sir Isaac Newton worked out a formula that, given a value close to a root, can move it closer.

Knowledge Box 3.2

Newton's method for finding roots

If x_0 is close to a root of $f(x)$, then the sequence generated by using the formula
$$x_{i+1} = x_i - \frac{f(x_i)}{f'(x_i)} \; \text{will approach the nearby root.}$$

Example 3.4   Use Newton's method with an initial guess of $x_0 = 1$ to approximate a root of $f(x) = x^2 - 2$.

Solution:

To start with, we need to find the Newton's method formula. Since $f'(x) = 2x$ we get that

$$x_{i+1} = x_i - \frac{x_i^2 - 2}{2x_i}$$

Now compute

$$x_1 = 1 - \frac{1 - 2}{2} = 1.5$$

$$x_2 = 1.5 - \frac{2.25 - 2}{3} \cong 1.4166666667$$

$$x_3 \cong 1.4142157$$

$$x_4 \cong 1.4142126$$

$$x_5 \cong 1.4142126$$

So by the fifth updating the approximation has stabilized at a value that agrees with $\sqrt{2}$ up to seven decimals. Since the roots of $f(x) = x^2 - 2$ are $\pm\sqrt{2}$ this is a nice demonstration that the technique works. Although we are demonstrating Newton's method on polynomials, it will work for any function that is continuous and differentiable near a root we are trying to find.

Newton's method – or other more sophisticated root finding methods – are often built into a calculator. By showing the values for x_1, x_2, and so on, coding the formula into a spreadsheet is a relatively low work method of doing the calculations. Once you code the formula you can just let the spreadsheet perform the iterations.

Example 3.5   Find the Newton's method formula for approximating roots of:

$$f(x) = x^3 + 3x - 1$$

Solution:

Just plug into the form for the Newton's method formula:

$$x_{i+1} = x_i - \frac{x_i^3 + 3x_i - 1}{3x_i^2 + 3}$$

If we start with $x_0 = 1.0$ we get:

$x_0 = 1$

$x_1 = 0.5$

$x_2 = 0.36111111$

$x_3 = 0.32917086$

$x_4 = 0.32335786$

$x_5 = 0.32237954$

$x_6 = 0.32221744$

$x_7 = 0.32219065$

$x_8 = 0.32218623$

$x_9 = 0.32218549$

$x_{10} = 0.32218537$

$x_{11} = 0.32218535$

$x_{12} = 0.32218535$

At which point the number has stopped changing and so is the root to 8 decimals. You can verify that $f(x)$ has only one root by graphing it.

The picture on the cover of this book demonstrates how Newton's method can be used to make pretty pictures. It plots points on the complex plane with the real part of a number plotted on the x-axis, and the imaginary part on the y-axis. It uses the polynomial:

$$x^6 + 2.43 \cdot x^4 - 5.8644 \cdot x^2 - 10.4976$$

This polynomial has six roots. Points are colored based on which root they converge to when used as a starting guess for Newton's method.

PROBLEMS

Problem 3.6   Compute the following polynomial products.

1. $(x^3 + 2x^2 + x + 3) \times (x + 5)$

2. $(x^2 + 4x - 1) \times (x^2 - 4x + 1)$

3. $(x^3 + 6x^2 + 11x + 6) \times (x^3 - 6x^2 + 11x - 6)$

4. $(x^3 + 3x + 7) \times (x^4 + x + 2)$

5. $(x + 1)^3 \times (x + 2)^3$

6. $(x^2 + x + 1)^3$

Problem 3.7   Perform the following polynomial divisions.

1. $(x^4 + 4x^3 + 6x^2 + 4x + 1) \div (x + 1)$

2. $(x^4 - x^3 + 2x^2 + x + 3) \div (x^2 + x + 1)$

3. $x^4 + x^3 + 2x^2 + x + 1) \div (x^2 + 1)$

4. $(x^5 + 8x^4 + 21x^3 + 35x^2 + 28x + 15) \div (x^2 + 2x + 3)$

5. $(x^6 - x^5 + 2x^4 - x^3 + 2x^2 - x + 1) \div (x^2 - x + 1)$

6. $(x^6 - 1) \div (x - 1)$

Problem 3.8   Use Newton's method to approximate the roots of the polynomial

$$f(x) = x^3 - 3x + 1$$

using initial guesses of $x_0 = -2, 0$, and 2. These should generate three distinct roots.

Problem 3.9   Use Newton's method to find a root of:

$$g(x) = x^5 - 5x^2 - 6$$

Problem 3.10   Earlier in the text it was asserted that a polynomial of odd degree must have at least one root. Explain why.

Problem 3.11   The polynomials

$$f(x) = x^2 - 2x + 1$$

and

$$g(x) = xr - 3x + 2$$

both have a root at $x = 1$. Use a spreadsheet to apply Newton's method to both of these problems with an initial guess of $x_0 = 0.4$. Both these intial guesses should converge to $x = 1$ – what is different about the convergence to one for these two polynomials. Why? Pictures may help.

Problem 3.12   Find all real roots of the following polynomials.

1. $f(x) = x^4 - 2x^3 - 2x^2 - 2x - 3$

2. $g(x) = x^3 - 6x^2 + 11x - 6$

3. $h(x) = x^6 - 64$

4. $q(x) = x^4 - 5x^2 + 4$

5. $r(x) = x^4 - 8x^2 + 5x + 6$

6. $s(x) = x^3 - 6x^2 + 12x - 8$

Problem 3.13   Find a polynomial of the form

$$x^4 + ax^3 + bx^2 + xc + d$$

with none of $a, b, c,$ or d zero that has no real roots at all.

Problem 3.14   Suppose that $p(x)$ is a polynomial. How many roots does $p(x)^2 + 1$ have? Justify your answer with one or more sentences.

Problem 3.15   Consider the polynomial

$$q(x) = x^3 - 25x$$

First find its three roots. Then figure out how many roots $q(x)^2 - 1$ has.

Problem 3.16   Find a polynomial with roots at $x = 1, 2, 3, 4,$ and 5.

Problem 3.17   Find a polynomial with roots at $x = 1, 2, 3, 4,$ and 5 that does not take on any negative values.

Problem 3.18   Given the Newton's method formula for a polynomial:

$$x_{i+1} = \frac{2x_i^3 - x_i^2 + 4}{3x_i^2 - 2x_i - 4}$$

what is the polynomial?

Problem 3.19   If the Newton's method formula for a polynomial is

$$x_{i+1} = \frac{3x_i^4 + x_i^2 - 7}{4x_i^4 + 10x_i}$$

what is the polynomial?

3.2   QUALITATIVE PROPERTIES OF POLYNOMIALS

Much of the work we do in a calculus class consists of calculating things. Some of the examples and homework problems are chosen to demonstrate the power of instead *deducing* things. In this section we will learn a number of properties of polynomials that help us deduce things about polynomials. Our first task is understanding how the roots of a polynomial affect the shape of its graph.

Knowledge Box 3.3

Rolle's Theorem

Suppose a continuous, differentiable function $y = f(x)$ has roots at $x = a$ and at $x = b$. Then for some c, $a \leq c \leq b$, $f'(c) = 0$.

The best way to understand Rolle's theorem is with a pictorial example. The function in Figure 3.1 has a root at $x = -2$ and one at $x = 3$. At $c = -\frac{1}{3}$ it has a critical value and so a horizontal tangent line.

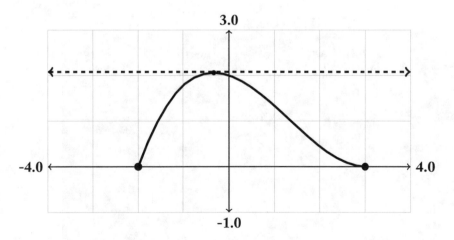

Figure 3.1: Illustration of Rolle's Theorem.

Colloquially, Rolle's theorem says there is (at least) one horizontal tangent between two roots. While we use it to describe polynomials, Rolle's theorem actually applies to all continuous differentiable functions.

The pictures in Figures 3.2 and 3.3 show examples of even and odd degree polynomials with the smallest possible number of roots.

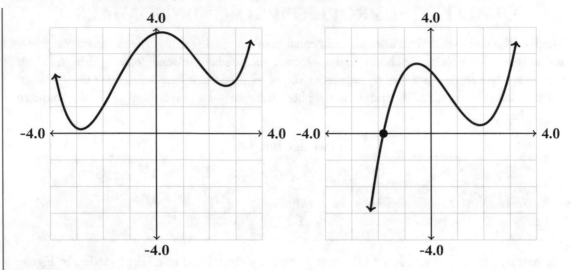

Figure 3.2: Fourth degree, no roots.   Figure 3.3: Third degree, one root.

Knowledge Box 3.4

The number of roots of a polynomial

- *An odd degree polynomial has at least one root.*

- *An even degree polynomial can have no roots.*

- *A polynomial of degree n has at most n roots.*

- *An exception to the previous rule is $f(x) = 0$ which has roots everywhere.*

- *Any number of roots between the minimum and maximum number of roots are possible.*

The fact that a polynomial of odd degree must have at least one root follows from the way its limits at infinity act. An odd degree polynomial must go to both of $\pm\infty$, which means that, on its way from one infinity to the other, it must pass through zero.

What does Rolle's theorem tell us about polynomials? Between any two roots there must be a horizontal tangent. Whenever there is a horizontal tangent, the function changes direction, so

Rolle's theorem tells us that the polynomial must change directions between roots.

A polynomial of degree n has *at most $n - 1$* horizontal tangents. This follows from the fact that the derivative has degree $n - 1$ and, so, at most $n - 1$ roots. The hill-tops and valley-bottoms of a polynomial correspond to the roots of its derivative.

Example 3.20   What is the smallest number of horizontal tangents a polynomial $p(x)$ may have?

Solution:

Since horizontal tangents are roots of $p'(x)$, the odd/even degree rules tell us that an odd degree polynomial may have *no* horizontal tangents, while an even degree polynomial must have at least *one*.

In fact, $p(x) = x^n$ exemplifies this minimum number of horizontal tangents. The even powers of x have a single critical value at $x = 0$ that yields a sign chart of

$$(+\infty) - - - (0) + + + (+\infty)$$

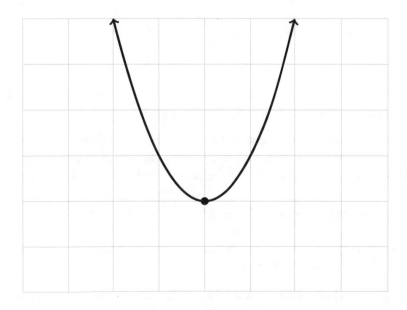

while the odd ones have a sign chart of

$$(-\infty) + + + (0) + + + (+\infty).$$

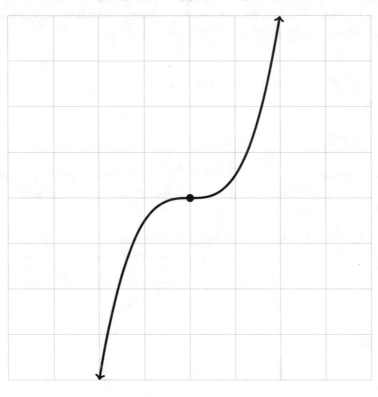

◇

3.2.1   MULTIPLICITY OF ROOTS

What makes it possible for a polynomial of degree n to have less than n roots? Part of this is we can add constants to a polynomial so that it intersects the x-axis as little as possible. Another factor is that roots can be repeated. Look at the polynomial shown in Figure 3.4:

$$p(x) = \frac{1}{4}(x + 2)(x - 2)^2 = \frac{1}{4}x^3 - \frac{1}{2}x^2 - x + 2$$

The fact that the root at $x = 2$ is squared makes the graph bounce off of the x axis instead of passing through. This leads to a definition.

Definition 3.1 *If* $(x - a)$ *divides a polynomial* $p(x)$, *then the highest power* k *so that* $(x - a)^k$ *divides* $p(x)$ *is the* **multiplicity** *of the root* $x = a$ *in* $p(x)$.

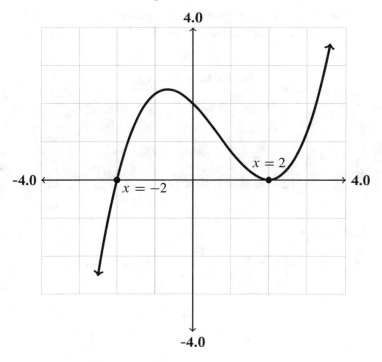

Figure 3.4: $p(x) = \dfrac{1}{4}(x+2)(x-2)^2$

The multiplicity of the root $x = 2$ in Figure 3.4 is two.

Knowledge Box 3.5

The effect of root multiplicity

If $p(x)$ is a polynomial, and $x = a$ is a root of $p(x)$, then

- *If the root at $x = a$ is of odd multiplicity, the graph of $p(x)$ passes through the x-axis at $x = a$.*

- *If the root at $x = a$ is of even multiplicity, the graph of $p(x)$ touches the x-axis and then bounces back on the same side it was on before at $x = a$.*

Notice that, since the root at $x = 2$ in Figure 3.4 has multiplicity two, there is a local minimum and a critical value at $x = 2$. This effect is fairly general and can be summarized as in Knowledge Box 3.6.

Knowledge Box 3.6

Roots with multiplicity and derivatives

If $p(x)$ is a polynomial, and $x = a$ is a root of $p(x)$ with multiplicity $k > 1$, then $x = a$ is also a root of $f'(x)$ with multiplicity $k - 1$.

Example 3.21   Verify that if

$$f(x) = (x - 2)^2(x + 2) = x^3 - 2x^2 - 4x + 8,$$

then $f'(2) = 0$.

Solution:

Compute

$$f'(x) = 3x^2 - 4x - 4$$

Then $f'(2) = 12 - 8 - 4 = 0$, and 2 is a root of the derivative.

The examples in this section show us how we can construct polynomials with particular properties quite easily by putting in the roots we want with the correct multiplicities. This sort of information is made even more useful by the fact that polynomials can be used to approximate other functions on bounded intervals.

PROBLEMS

Problem 3.22   Show that the number of horizontal tangents between two roots – that do not have another root between them – must be odd.

Problem 3.23   Construct a polynomial that has more than one horizontal tangent between two roots that do not have another root between them.

Problem 3.24   Prove that a non-constant polynomial can have zero inflection values; do this by giving an example of one.

Problem 3.25   Show that a polynomial of degree n can have at most $n - 2$ inflection values.

Problem 3.26   Give an example of a non-constant polynomial with two roots that never takes on negative values. Explain why your example is correct.

Problem 3.27   Suppose a polynomial has three distinct roots, r_1, r_2, r_3. What can you deduce from this fact alone about the degree of the polynomial?

Problem 3.28   Suppose we have two polynomials $p(x)$ and $q(x)$ and that all roots of $p(x)$ are roots of $q(x)$ but at least one root of $q(x)$ is not a root of $p(x)$. If $p(x)$ has two roots, provide three examples. In the first, $p(x)$ should have a degree less than $q(x)$; in the second, $p(x)$ and $q(x)$ have equal degree; in the third, $p(x)$ should have a higher degree than $q(x)$.

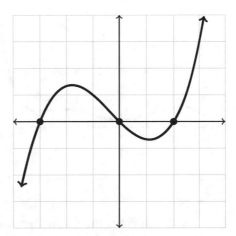

Problem 3.29   Deduce as much as you can from the graph above. The grids are of length one unit. In particular, what is the minimum degree and what can you deduce about the roots and their multiplicity?

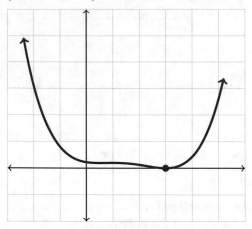

Problem 3.30   Deduce as much as you can from the graph above. The grids are of length one unit. What is the minimum possible degree and what can you deduce about the roots and their multiplicity?

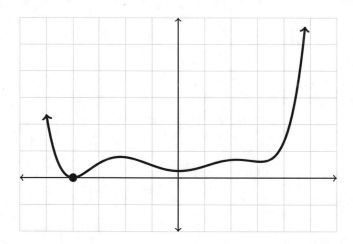

Problem 3.31   Deduce as much as you can from the graph above. The grids are of length one unit. In particular what is the minimum degree and what is possible to deduce about the roots and their multiplicity?

Problem 3.32   What is the smallest possible degree for a polynomial that is an answer to Question 3.26.

Problem 3.33   Using the product rule, prove the statement: If $p(x)$ is a polynomial and $x = a$ is a root of $p(x)$ with multiplicity $k > 1$, then $x = a$ is also a root of $f'(x)$ (from Knowledge Box 3.6).

Problem 3.34   Using material from this section, prove that

$$y = \cos(x)$$

is not a polynomial function.

Problem 3.35   Suppose two polynomials share the same roots and, for those roots, the same multiplicities. If the polynomials are not equal, how do they differ?

3.3   L'HÔPITAL'S RULE; STRANGE POLYNOMIALS

During our study of curve sketching, we developed a rule of thumb about computing the limit at infinity of the ratio of two polynomials. If $p(x) = ax^n + \cdots$ is a polynomial of degree n and $q(x) = bx^m + \cdots$ is a polynomial of degree m the rule said:

$$\lim_{x \to \infty} \frac{p(x)}{q(x)} = \begin{cases} \pm\infty & n > m \\ a/b & n = m \\ 0 & n < m \end{cases}$$

This rule is a consequence of **L'Hôpital's rule**, a useful tool for computing limits that cannot be resolved other ways.

Knowledge Box 3.7

L'Hôpital's rule at $x = c$

If $\lim_{x \to c} f(x) = 0$ *and* $\lim_{x \to c} g(x) = 0$

or

if $\lim_{x \to c} f(x) = \pm\infty$ *and* $\lim_{x \to c} g(x) = \pm\infty$, *then*

$$\lim_{x \to c} \frac{f(x)}{g(x)} = \lim_{x \to c} \frac{f'(x)}{g'(x)}$$

The hypotheses that the limits of the numerator and denominator are both going to zero or both going to some sort of infinity are critical. L'Hôpital's rule will give the wrong answer if the hypotheses are not satisfied.

Let's do an example.

Example 3.36   Compute

$$\lim_{x \to 2} \frac{x^2 - 4}{x - 2}$$

with L'Hôpital's rule.

Solution:

We could resolve this limit by factoring and canceling, but, since the top and bottom of the fraction are both 0 at $x = 2$, L'Hôpital's rule applies.

$$\lim_{x \to 2} \frac{x^2 - 4}{x - 2} = \lim_{x \to 2} \frac{2x}{1} = \frac{4}{1} = 4$$

$\Diamond$

Knowledge Box 3.8

L'Hôpital's rule as $x \to \infty$

If $\lim_{x \to \infty} f(x) = 0$ and $\lim_{x \to \infty} g(x) = 0$
or
if $\lim_{x \to \infty} f(x) = \pm\infty$ and $\lim_{x \to \infty} g(x) = \pm\infty$, then

$$\lim_{x \to \infty} \frac{f(x)}{g(x)} = \lim_{x \to \infty} \frac{f'(x)}{g'(x)}$$

This version of L'Hôpital's rule is the one that gives us the rule-of-thumb for the limit at infinity of ratios of polynomials.

Let's do an example.

Example 3.37   Compute

$$\lim_{x \to \infty} \frac{x^2 + 2x + 3}{3x^2 + 1}.$$

Solution:

Both the numerator and denominator are going to infinity, so:

$$\lim_{x \to \infty} \frac{x^2 + 2x + 3}{3x^2 + 1} = \lim_{x \to \infty} \frac{2x + 2}{6x} \qquad \text{L'Hôpital's rule still applies}$$

$$= \lim_{x \to \infty} \frac{2}{6}$$

$$= \frac{1}{3}$$

And so the limit at infinity is $\frac{1}{3}$ – exactly what the rule of thumb gives us.

The reason the rule of thumb works for polynomials of equal degree is that L'Hôpital's rule applies until we have taken so many derivatives that the numerator and denominator are constants, at which point the accumulated values from bringing the power out front cancel, leaving the ratio of the highest degree coefficients as the limit.

At this point, we add a name to our mathematical vocabulary with the following definition.

Definition 3.2 *A **rational function** is a function that is the ratio of polynomial functions.*

We have been working with and graphing rational functions for some time now, so they are not new. We just have a name for them now. While L'Hôpital's rule was introduced to back-fill the rule of thumb for limits at infinity of rational functions (to use their new name), it in fact applies to all continuous, differentiable functions.

Let's do a couple of examples.

Example 3.38   Compute

$$\lim_{x \to 0} \frac{\sin(x)}{x}.$$

Solution:

The numerator and denominator both go to zero, so L'Hôpital's rule applies.

$$\lim_{x \to 0} \frac{\sin(x)}{x} = \lim_{x \to 0} \frac{\cos(x)}{1} = 1$$

$\Diamond$

Example 3.39   Compute

$$\lim_{x \to 0} \frac{\cos(x) - 1}{x}.$$

Solution:

The numerator and denominator are both going to zero, so L'Hôpital's rule applies.

$$\lim_{x \to 0} \frac{\cos(x) - 1}{x} = \lim_{x \to 0} \frac{-\sin(x)}{1} = 0$$

$\Diamond$

As with every other technique we have learned, we may use algebraic rearrangement to permit us to extend the reach of L'Hôpital's rule.

Example 3.40   Compute

$$\lim_{x \to 0} x \cdot \ln(x).$$

Solution:

This problem does not fit the form for L'Hôpital's rule, but we can coerce it into that form:

$$\lim_{x \to 0} x \cdot \ln(x) = \lim_{x \to 0} \frac{\ln(x)}{\dfrac{1}{x}} \qquad \text{Now in correct form } \frac{-\infty}{\infty} \text{ for L'Hôpital}$$

$$= \lim_{x \to 0} \frac{\dfrac{1}{x}}{\dfrac{-1}{x^2}} \qquad \text{Use L'Hôpital}$$

$$= \lim_{x \to 0} -\frac{x^2}{x} = \lim_{x \to 0} -\frac{x}{1} \qquad \text{Simplify}$$

$$= 0 \qquad \text{Done}$$

So x moves toward zero faster than $\ln(x)$ moves toward negative infinity as $x \to 0$.

$$\Diamond$$

It is possible for L'Hôpital to take you to a limit that clearly does not exist. As before, the rules-of-thumb developed for curve sketching can also supply the answer, but in mathematics a solid method is preferred.

Example 3.41   Find $\lim\limits_{x \to \infty} \dfrac{x^3 + 1}{x^2}$.

Solution:

The rule-of-thumb tells us this diverges to ∞, but let's work through this with L'Hôpital's rule and see where it goes.

$$\lim_{x \to \infty} \frac{x^3 + 1}{x^2} = \lim_{x \to \infty} \frac{3x^2}{2x} = \lim_{x \to \infty} \frac{3}{2}x \to \infty$$

L'Hôpital's rules also tells us this diverges to infinity.

$$\Diamond$$

Since this chapter is about polynomials, Knowledge Box 3.9 gives a number of useful polynomial identities that did not really fit before this point.

Knowledge Box 3.9

Useful polynomial identities

- $x^2 - a^2 = (x - a)(x + a)$
- $x^3 - a^3 = (x - a)(x^2 + ax + a^2)$
- $x^3 + a^3 = (x + a)(x^2 - ax + a^2)$
- $x^n - a^n = (x - a)(x^{n-1} + ax^{n-2} + \cdots a^{n-2}x + a^{n-1})$

That last identity is quite useful for adding up powers of a variable. Another form of the identity, with $a = 1$ is

$$\frac{x^n - 1}{x - 1} = x^{n-1} + x^{n-2} + \cdots + x + 1$$

How can this be applied?

Example 3.42   Compute

$$\sum_{k=0}^{20} 1.5^k$$

Solution:

$$\sum_{k=0}^{20} 1.5^k = \frac{1.5^{21} - 1}{1.5 - 1} = \frac{\left(\frac{3}{2}\right)^{21} - 1}{\frac{1}{2}} = \frac{3^{21} - 2^{21}}{2^{20}} = \frac{10458256051}{1048576} \cong 9973.8$$

◊

Let's do another with slightly less insane numbers.

Example 3.43   Find

$$\sum_{k=0}^{11} 3^k$$

Solution:

$$\sum_{k=0}^{11} 3^k = \frac{3^{12} - 1}{3 - 1} = \frac{1}{2} \cdot 531440 = 265720$$

$\Diamond$

This material is very important later, when we study sequences and series.

3.3.1 STRANGE POLYNOMIALS

The goal of this section is to let us broaden the reach of what we know about polynomials to other functions. Suppose that we wish to solve the equation:

$$e^{2x} - 3e^x + 2 = 0$$

This is clearly not a polynomial equation, but we can find a polynomial with the following trick. Let $u = e^x$. Then, since $e^{2x} = (e^x)^2 = u^2$, we transform the original problem into:

$$u^2 - 3u + 2 = 0$$

This factors into $(u - 2)(u - 1) = 0$ so $u = 1, 2$. Reversing the transformation $e^x = 1, 2$. Take the log of both sides and the result becomes $x = \ln(1), \ln(2)$ or $x = 0, \ln(2)$.

This trick, called u-substitution, lets us turn one type of equation into another. This technique will become *very* useful in Chapter 4, but for now it permits us to solve a wider variety of equations. Some care is required.

Example 3.44   Find the solutions to the equation:

$$2\sin^2(x) - 5\sin(x) + 2$$

Solution:

Try $u = \sin(x)$ changing the problem to:

$$2u^2 - 5u + 2 = 0$$

As in the previous example, this factors, giving us $(2u - 1)(u - 2) = 0$, and we get that $u = 1/2, 2$.

So far, so good.

But when we reverse the transformation we get $\sin(x) = 1/2, 2$ and $\sin(x) = 2$ is impossible! We know that $-1 \leq \sin(x) \leq 1$.

So even though there are two values for u, only one leads to a solution of the original equation.

$$\sin(x) = \frac{1}{2} \quad \text{or} \quad x = \frac{\pi}{6}$$

Giving the sole solution to the problem.

Example 3.45   Solve $\ln(x)^2 - 3\ln(x) - 3 = 0$.

Solution:

Here the obvious substitution is $u = \ln(x)$, and we get $u^2 - 3u - 3 = 0$. Apply the quadratic equation and reverse the transformation:

$$u = \frac{3 \pm \sqrt{9 - 4(1)(-3)}}{2}$$

$$= \frac{3 \pm \sqrt{21}}{2} \qquad \qquad \text{Only one positive root}$$

$$\ln(x) = \frac{3 + \sqrt{21}}{2} \cong 3.79$$

$$x \cong 1.33$$

We dismissed the negative root because only positive numbers have logs.

This trick sometimes helps with polynomials as well.

Example 3.46   Solve $x^4 - 6x^2 + 8 = 0$.

Solution:

Take $u = x^2$. Then

$$u^2 - 6u + 8 = 0$$

Factor: $(u-2)(u-4) = 0$, and we get $u = 2, 4$. So $x^2 = 2, 4$ and thus $x = \pm\sqrt{2}, \pm 2$.

PROBLEMS

Problem 3.47   For each of the following limits, use L'Hôpital's rule to resolve the limit.

1. $\displaystyle\lim_{x\to 3} \frac{x-3}{x^2-9}$

2. $\displaystyle\lim_{x\to 2} \frac{x^3-x-6}{x^2-4}$

3. $\displaystyle\lim_{x\to 1} \frac{x^6-1}{x^2-1}$

4. $\displaystyle\lim_{x\to -2} \frac{x^3+8}{x^3+x^2-x+2}$

5. $\displaystyle\lim_{x\to 2} \frac{x^5-32}{x^2-4}$

6. $\displaystyle\lim_{x\to \pi} \frac{\cos(x)+1}{x-\pi}$

Problem 3.48   Before, when we had to resolve a limit of the form $\frac{0}{0}$ by algebra using cancellation, we had to know the factorization of the numerator and denominator.

Discuss: does L'Hôpital's rule make this process easier?

Problem 3.49   Find, in general, using L'-Hôpital's rule

$$\lim_{x\to 1} \frac{x^n-1}{x^2-1}$$

Problem 3.50   For each of the following limits, use L'Hôpital's rule to resolve the limit.

1. $\lim\limits_{x\to\infty} \dfrac{x^2 + 3x + 5}{x^2 - 3x + 4}$

2. $\lim\limits_{x\to\infty} \dfrac{2e^x}{e^x + 4}$

3. $\lim\limits_{x\to\infty} \dfrac{x + e^x}{x + x^2 + 3e^x}$

4. $\lim\limits_{x\to 1} (x - 1)\cdot \ln(x^3 - 1)$

5. $\lim\limits_{x\to 0} \dfrac{\sin(3x)}{2x}$

6. $\lim\limits_{x\to 0} \dfrac{\cos(x) - 1}{x^2}$

Problem 3.51   Compute the following sums with a polynomial identity. Some of these may require using the identity twice.

1. $\sum\limits_{k=0}^{19} 0.9^k$

2. $\sum\limits_{k=0}^{25} 1.2^k$

3. $\sum\limits_{k=0}^{8} 7^k$

4. $\sum\limits_{k=5}^{15} 1.5^k$

5. $\sum\limits_{k=0}^{12} 0.1^k$

6. $\sum\limits_{k=0}^{10} (-0.5)^k$

Problem 3.52   Give an example of a polynomial $p(x)$ with two roots for which

$$p(\sin(x)) = 0$$

has no solutions. Explain why your solution is correct.

Problem 3.53   If $p(x)$ is a polynomial with k roots, what are the possible number of solutions to

$$p(\ln(x)) = 0?$$

Problem 3.54   Solve the following equations. Note that they are, in a sense, polynomials.

1. $2\sin^2(x) + \sin(x) = 0$

2. $4\cos^2(x) - 3 = 0$

3. $4\sin^3(x) - \sin(x) = 0$

4. $e^{2x} - 7e^x + 12 = 0$

5. $e^{3x} - 6e^{2x} + 11e^x - 6 = 0$

6. $\ln^2(x) - 8\ln(x) + 15 = 0$

7. $\cos^2(x) - \sin^2(x) - \cos^2(2x) = 0$

8. $\tan^3(x) - \tan(x) = 0$

Problem 3.55   Factor $x^6 - 729$ as far as you can.

Problem 3.56   Factor $x^8 - 256$ as far as you can.

Problem 3.57   If we know that $p(x)$ is a polynomial with n roots and $q(x)$ is a polynomial with m roots, with $n \leq m$, then what do we know about the number of roots of

$$h(x) = p(x) \cdot q(x)?$$

CHAPTER 4

Methods of Integration I

The integration methods we have learned thus far are based on the fact that integrals are reversed derivatives. To some extent we can increase the reach of the "reverse derivative" technique by setting things up with algebra. The chapters on methods of integration will introduce the most useful and broadly applicable of thousands of integration methods that have been found over the last several centuries.

4.1 u-SUBSTITUTION

Every integral method is a reversed derivative, and u-substitution is the reverse of the chain rule. We introduced u-substitution in Chapter 3 to permit us to solve more complex equations with polynomial techniques. For integrals the technique is much the same, except that we need to worry about the differential. Let's start with an example.

Example 4.1   Find $\int 2xe^{x^2} \cdot dx$.

Solution:

This is not an integral for which we already have a form. Set $u = x^2$ and then compute $du = 2x \cdot dx$. These pieces make up all of the integral and we can now solve the problem as follows:

$$\int 2xe^{x^2} \cdot dx = \int e^{x^2} \cdot (2x \cdot dx) \qquad \text{Set it up}$$

$$= \int e^u \cdot du \qquad \text{Substitute}$$

$$= e^u + C \qquad \text{Integrate}$$

$$= e^{x^2} + C \qquad \text{At the end, reverse the substitution.}$$

$\Diamond$

Recall that you can check an integral by taking a derivative, so

$$\left(e^{x^2} + C\right)' = e^{x^2} \cdot 2x = 2xe^{x^2}$$

and we see the substitution worked to let us do this integral. Notice that the chain rule re-created *du* for us.

Let's do some more examples.

Example 4.2   Find $\int \cos(2x) \cdot dx$.

Solution:

We know that the integral of cosine is sine. So the only problem is the $2x$. This is where to make our *u*-substitution.

If we let $u = 2x$ then $du = 2dx$ so $dx = \dfrac{1}{2}du$.

Now we substitute and integrate:

$$\int \cos(2x) \cdot dx = \int \cos(u)\frac{1}{2}du$$

$$= \frac{1}{2}\int \cos(u) \cdot du$$

$$= \frac{1}{2}\sin(u) + C$$

$$= \frac{1}{2}\sin(2x) + C$$

This last example has a very general version that can let you do some types of *u*-substitution in your head.

Example 4.3   Suppose that $F(x) + C = \displaystyle\int f(x) \cdot dx$. Then what is $\displaystyle\int f(ax + b) \cdot dx$?

Solution:

Let $u = ax + b$ and so $du = a \cdot dx$ and $dx = \dfrac{1}{a} \cdot du$. Then:

$$\int f(ax + b) \cdot dx = \int f(u) \cdot \frac{1}{a} \cdot du$$

$$= \frac{1}{a} \int f(u) \cdot du$$

$$= \frac{1}{a} F(u) + C$$

$$= \frac{1}{a} F(ax + b) + C$$

$\Diamond$

This is important enough to put in a Knowledge Box. What we are doing is correcting for composing a function with a linear function.

Knowledge Box 4.1

u-substitution to correct a linear composition

If $\displaystyle\int f(x) \cdot dx = F(x) + C$, *then*

$$\int f(ax + b) \cdot dx = \frac{1}{a} F(ax + b) + C$$

Example 4.4   Find $\displaystyle\int e^{0.5x - 1} dx$.

Solution:

$$2e^{0.5x - 1} + C$$

by directly applying the linear correction with $a = 0.5$ and $b = -1$.

$\Diamond$

This example shows how much calculation the linear correction technique saves. Sometimes a u substitution is not obvious.

Example 4.5   Find

$$\int \frac{dx}{x \ln(x)}$$

Solution:

The key to this one is remembering that $\dfrac{d}{dx} \ln(x) = \dfrac{1}{x}$ making $u = \ln(x)$ a natural choice to try. Then $du = \dfrac{dx}{x}$ and we can integrate.

$$\int \frac{dx}{x \ln(x)} = \int \frac{1}{\ln(x)} \cdot \frac{dx}{x}$$

$$= \int \frac{1}{u} \cdot du$$

$$= \ln(u) + C$$

$$= \ln(\ln(x)) + C$$

$$\Diamond$$

Sometimes the substitution is so unobvious that people clearly figured out the integral another way. The following example has this property.

Example 4.6   Find

$$\int \sec(x) \cdot dx.$$

Solution:

We begin with an algebraic transformation.

$$\int \sec(x) \cdot dx = \int \sec(x) \frac{\sec(x) + \tan(x)}{\sec(x) + \tan(x)} \cdot dx$$

$$= \int \frac{\sec^2(x) + \sec(x) \tan(x)}{\tan(x) + \sec(x)} \cdot dx$$

At this point let $u = \tan(x) + \sec(x)$ so that

$$du = \left(\sec^2(x) + \sec(x)\tan(x)\right)dx$$

which makes this all:

$$\int \sec(x) \cdot dx = \int \frac{du}{u}$$

$$= \ln(|u|) + C$$

$$= \ln(|\tan(x) + \sec(x)|) + C$$

The absolute value bars are needed because the value of these functions can be negative.

◊

We also now know enough to capture the integral of another of the basic functions.

Example 4.7   Compute $\int \tan(x) \cdot dx$.

Solution:

Since $\tan(x) = \dfrac{\sin(x)}{\cos(x)}$ we let $u = \cos(x)$ obtaining $du = -\sin(x)dx$ or $-du = \sin(x)dx$. Plugging in the substitution we get

$$\int \tan(x) \cdot dx = \int \frac{-du}{u} = -\int \frac{du}{u}$$

$$= -\ln(|u|) + C$$

$$= \ln\left(\frac{1}{|u|}\right) + C$$

$$= \ln\left(\frac{1}{|\cos(x)|}\right) + C$$

$$= \ln(|\sec(x)|) + C$$

Taking the minus sign inside the log is a negative first power or reciprocal, which is how we get to the secant.

◊

Knowledge Box 4.2

Basic trig integrals from u-substitution

- $\int \tan(x) \cdot dx = \ln(|\sec(x)|) + C$

- $\int \sec(x) \cdot dx = \ln(|\sec(x) + \tan(x)|) + C$

It is possible to have fairly obvious u-substitution problems.

Example 4.8   Find

$$\int x(x^2 + 3)^6 \cdot dx.$$

Solution:

Let $u = x^2 + 3$; then $du = 2x \cdot dx$ and so $\dfrac{1}{2}du = x \cdot dx$. Integrate:

$$\int x(x^2 + 3)^6 \cdot dx = \int (x^2 + 3)^6 \cdot (x \cdot dx)$$

$$= \int u^6 \frac{1}{2}du$$

$$= \frac{1}{2}\int u^6 \cdot du$$

$$= \frac{1}{2} \times \frac{1}{7}u^7 + C$$

$$= \frac{1}{14}(x^2 + 3)^7 + C$$

$$\Diamond$$

It is possible to need algebra to set up a *u*-substitution, often in less apocalyptic ways than the integral of secant, earlier in this section.

Example 4.9   Compute

$$\int \frac{x^2 + 2x + 1}{x^2 + 1} \cdot dx$$

Solution:

$$\int \frac{x^2 + 2x + 1}{x^2 + 1} \cdot dx = \int \frac{x^2 + 1}{x^2 + 1} dx + \int \frac{2x}{x^2 + 1} dx$$

$$= \int dx + \int \frac{2x}{x^2 + 1} dx$$

$$= x + \int \frac{2x}{x^2 + 1} dx + C$$

At this point let $u = x^2 + 1$ so that $du = 2x \cdot dx$ and finish

$$\int \frac{x^2 + 2x + 1}{x^2 + 1} \cdot dx = x + \int \frac{du}{u} + C$$

$$= x + \ln(u) + C$$

$$= x + \ln(x^2 + 1) + C$$

Done! If you have a rational function whose denominator is not higher degree, dividing first is a good way to simplify things.

Another use of u-substitution is to integrate odd powers of sine and cosine.

Example 4.10   Find

$$\int \cos^3(x) \cdot dx.$$

Solution:

$$\int \cos^3(x) \cdot dx = \int \cos^2(x) \cdot \cos(x) \cdot dx$$

$$= \int (1 - \sin^2(x)) \cdot \cos(x) \cdot dx$$

$$= \int \cos(x) \cdot dx - \int \sin^2(x) \cos(x) \cdot dx$$

$$= \sin(x) - \int \sin^2(x) \cos(x) \cdot dx$$

Let $u = \sin(x)$ so that $du = \cos(x) \cdot dx$

$$= \sin(x) - \int u^2 \cdot du$$

$$= \sin(x) - \frac{1}{3}u^3 + C$$

$$= \sin(x) - \frac{1}{3}\sin^3(x) + C$$

◊

We have enough machinery to integrate even powers of sine and cosine now. We will learn how to integrate odd powers in Section 4.3.

Example 4.11   Find

$$\int \sin^2(x) \cdot dx.$$

Solution:

$$\int \sin^2(x) \cdot dx = \int \frac{1}{2}(1 - \cos(2x)) \cdot dx = \frac{1}{2}(x - \frac{1}{2}\sin(2x)) + C$$

We are using a trig identity for the second step and the answer is $\frac{1}{2}x - \frac{1}{4}\sin(2x) + C$.

4.1.1   SUBSTITUTION IN DEFINITE INTEGRALS

An issue we have not dealt with is substitution in definite integrals. There are two viable approaches to this topic.

- Substitute back to the original variable before plugging in the limits.

- Apply the substitution to the limits to get new limits.

So far we have always substituted back to the original variable to finish the integral.

Example 4.12   Compute

$$\int_0^3 x\sqrt{x^2 + 4} \cdot dx$$

Solution:

This is a simple *u*-substitution. Let $u = x^2 + 4$, so that $du = 2x \cdot dx$ and $\frac{1}{2}du = x \cdot dx$. Applying the substitution to the limits we get that 0 to 3 becomes $u(0) = 4$ and $u(3) = 13$.

$$\int_0^3 x\sqrt{x^2 + 4} \cdot dx = \int_4^{13} \sqrt{u} \cdot \frac{1}{2} du$$

$$= \frac{1}{2} \int_4^{13} u^{1/2} \cdot du$$

$$= \frac{1}{2} \cdot \frac{2}{3} u^{3/2} \Big|_4^{13}$$

$$= \frac{1}{3} \left(13^{3/2} - 4^{3/2} \right)$$

$$= \frac{1}{3} \left(\sqrt{2197} - 8 \right) = \cong 12.96$$

$\Diamond$

Example 4.13   Compute

$$\int_0^1 \frac{e^x}{e^{2x} + 1} \cdot dx$$

Solution:

Let $u = e^x$ so that $du = e^x \cdot dx$. The limits become $u(0) = 1$ and $u(1) = e$.

$$\int_0^1 \frac{e^x}{e^{2x} + 1} \cdot dx = \int_1^e \frac{du}{u^2 + 1}$$

$$= \tan^{-1}(u) \Big|_1^e$$

$$= \tan^{-1}(e) - \tan^{-1}(0)$$

$$= \tan^{-1}(e)$$

$$\cong 1.218$$

$\Diamond$

PROBLEMS

Problem 4.14   Integrate the following by using u-substitution. Explicitly state u.

1. $\int \sin(3x) \cdot dx$

2. $\int \sec^2(5x + 1) \cdot dx$

3. $\int_0^4 \frac{x}{x^2 + 1} \cdot dx$

4. $\int_1^3 4x \cdot (2x^2 + 1)^5 \cdot dx$

5. $\int_1^4 \frac{1}{x \cdot \ln^2(x)} \cdot dx$

6. $\int (\sin(x) + \cos(x))^2 \cdot dx$

Problem 4.15   Integrate the following.

1. $\int \cos(\pi x) \cdot dx$

2. $\int e^{3x+2} \cdot dx$

3. $\int \sin(14x - 5) \cdot dx$

4. $\int_0^1 (3x + 4)^{12} \cdot dx$

5. $\int \sec(2x) \tan(2x) \cdot dx$

6. $\int_0^1 \frac{dx}{3x + 5}$

Problem 4.16   Show that if $F(x) + C = \int f(x) \cdot dx$, then $\int x \cdot f(x^2 + a) \cdot dx = \frac{1}{2} F(x^2 + a) + C$.

Problem 4.17   Compute $\int \cot(x) \cdot dx$.

Problem 4.18   Compute $\int \csc(x) \cdot dx$.

Problem 4.19   Compute $\int_0^{\pi/4} \sin^4(x) \cos(x) \cdot dx$.

Problem 4.20   Compute $\int_0^{\pi/2} \cos^4(x) \cdot dx$.

Problem 4.21   Compute $\int \cos^5(x) \cdot dx$.

Problem 4.22   Compute $\int \sin^3(x) \cos^3(x) \cdot dx$.

Problem 4.23   Compute $\int \tan^3(x) \sec^2(x) \cdot dx$.

Problem 4.24   Compute $\int \frac{\sin^3(x)}{\cos^5(x)} \cdot dx$.

Problem 4.25   Integrate the following by using u-substitution, using algebra as needed. Explicitly state u.

1. $\int x e^{x^2} \cdot dx$

2. $\int_0^1 \frac{x^2 + 4x + 2}{x^2 + 2} \cdot dx$

3. $\int \frac{x^5}{x^2 + 1} \cdot dx$

4. $\int \frac{\ln^2(x) + 4\ln(x) + 2}{x} \cdot dx$

5. $\int \frac{\sin^2(x)}{\cos^2(x) + 1} \cdot dx$

6. $\int_0^2 \frac{2x + 1}{(x^2 + x + 1)^5} \cdot dx$

7. $\int \cos^7(2x) \sin(x) \cos(x) \cdot dx$

8. $\int \frac{e^x}{e^x + 1} \cdot dx$

9. $\int \frac{\tan^{-1}(x)}{x^2 + 1} \cdot dx$

Problem 4.26   Perform the following definite integrals.

1. $\int_0^4 (2x - 3)^8 \cdot dx$

2. $\int_0^{\sqrt{\pi}} x \sin(x^2) \cdot dx$

3. $\int_0^1 (3x + 1)e^{9x^2+6x+1} \cdot dx$

4. $\int_1^2 \frac{\ln(x)}{x} \cdot dx$

5. $\int_{-2}^2 \frac{x^5}{x^2 + 1} \cdot dx$

6. $\int_0^{\sqrt[4]{3}} \frac{2x}{x^4 + 1} \cdot dx$

4.2   INTEGRATION BY PARTS

Probably the easiest of the complex derivative rules is the product rule. The reversed product rule is a technique called **integration by parts**. Let's derive it.

$$(UV)' = V \cdot dU + U \cdot dV \qquad \text{This is just the product rule}$$

$$U \cdot dV = (UV)' - V \cdot dU \qquad \text{Rearrange}$$

$$\int U \cdot dV = \int (UV)' - \int V \cdot dU \qquad \text{Integrate both sides}$$

$$\int U \cdot dV = UV - \int V \cdot dU \qquad \text{This is the formula}$$

Knowledge Box 4.3

Integration by parts

$$\int U \cdot dV = UV - \int V \cdot dU$$

The technique for using this rule is not obvious from its statement. The "parts" are U and V together with their derivatives. Examples are needed.

Example 4.27   Compute

$$\int x \cdot \cos(x) \cdot dx.$$

Solution:

Choose $U = x$ and $dV = \cos(x) \cdot dx$.

Then $dU = dx$ and $V = \int \cos(x) \cdot dx = \sin(x)$.

When doing integration by parts we add the $+C$ as the last step of the integration process.

Now that we have the parts and their derivatives, we apply the integration by parts formula.

$$\int U \cdot dV = UV - \int V \cdot dU$$

$$\int x \cdot \cos(x) \cdot dx = x \cdot \sin(x) - \int \sin(x) \cdot dx$$

$$= x \cdot \sin(x) - (-\cos(x)) + C$$

$$= x \cdot \sin(x) + \cos(x) + C$$

Let's check this result by taking the derivative:

$$(x \cdot \sin(x) + \cos(x) + C)' = \sin(x) + x \cdot \cos(x) - \sin(x) = x \cdot \cos(x)$$

So the method worked.

$\Diamond$

There is a substantial strategic component to choosing the parts U and dV when doing integration by parts. You take the derivative of U, and you must integrate dV, and when you're done it would be lovely if the result could be integrated without too much difficulty. In general you choose U so that differentiation will make a problem go away.

Let's do another simple example.

Example 4.28   Find

$$\int x \, e^x \cdot dx.$$

Solution:

We can integrate e^x, and x goes away if we take its derivative.

So, choose $U = x$, $dV = e^x \cdot dx$. This means $dU = dx$ and $V = e^x$.

Apply the formula:

$$\int U \cdot dV = UV - \int V \cdot dU$$

$$\int xe^x \cdot dx = xe^x - \int e^x \cdot dx$$

$$= xe^x - e^x + C$$

$$\Diamond$$

Sometimes the choice of parts is not obvious. Let's pick up another basic integral using integration by parts.

Example 4.29   Compute

$$\int \tan^{-1}(x) \cdot dx$$

Solution:

At first it might look like there are no parts, but there are.

Choose $U = \tan^{-1}(x)$ and $dV = dx$. Then $dU = \dfrac{dx}{x^2 + 1}$ and $V = x$.

Apply the formula:

$$\int U \cdot dV = UV - \int V \cdot dU$$

$$\int \tan^{-1}(x) \cdot dx = x \cdot \tan^{-1}(x) - \int \frac{x}{x^2 + 1} \cdot dx \qquad \text{This is a substitution integral.}$$

Let $r = x^2 + 1$. Then $\frac{1}{2}dr = x \cdot dx$.

$$= x \cdot \tan^{-1}(x) - \frac{1}{2}\int \frac{dr}{r}$$

$$= x \cdot \tan^{-1}(x) - \frac{1}{2}\ln(r) + C$$

$$= x \cdot \tan^{-1}(x) - \frac{1}{2}\ln(x^2 + 1) + C$$

Notice that, since U and dU are used in integration by parts, we used "r-substitution" instead of u-substitution for the part of the integral that needed substitution.

Knowledge Box 4.4

Integral of the arctangent function

$$\int \tan^{-1}(x) \cdot dx = x \cdot \tan^{-1}(x) - \frac{1}{2}\ln(x^2 + 1) + C$$

Example 4.30   Compute $\int x^2 \cdot \ln(x) \cdot dx$.

Solution:

Tactically, we know that $\ln(x)$ turns into a (negative) power of x when we take its derivative. So a natural choice to try is $U = \ln(x)$, $dV = x^2 \cdot dx$.

This makes $dU = \frac{dx}{x}$ and $V = \frac{1}{3}x^3$.

Apply the formula.

$$\int U \cdot dV = UV - \int V \cdot dU$$

$$\int x^2 \cdot \ln(x) \cdot dx = \frac{1}{3}x^3 \cdot \ln(x) - \int \frac{1}{3}x^3 \cdot \frac{dx}{x}$$

$$= \frac{1}{3}x^3 \cdot \ln(x) - \frac{1}{3}\int x^2 \cdot dx$$

$$= \frac{1}{3}x^3 \cdot \ln(x) - \frac{1}{9}x^3 + C$$

$$\Diamond$$

Sometimes integration by parts must be applied more than once to finish a problem.

Example 4.31   Compute $\int x^2 \cdot e^x \cdot dx$.

Solution:

Since taking the derivative of a power of x at least makes it a smaller power of x, choose $u = x^2, dV = e^x \cdot dx$.

Then $dU = 2x \cdot dx$ and $V = e^x$.

Integrate by parts:

$$\int x^2 \cdot e^x \cdot dx = x^2 \cdot e^x - \int 2x \cdot e^x \cdot dx$$

$$= x^2 \cdot e^x - 2\int x \cdot e^x \cdot dx \qquad \text{Integrate by parts again.}$$

$$U = x; \ dV = e^x \cdot dx$$

$$dU = dx; \ V = e^x$$

$$= x^2 \cdot e^x - 2\left(x \cdot e^x - \int e^x \cdot dx\right)$$

$$= x^2 \cdot e^x - 2x \cdot e^x + 2e^x + C$$

$$\Diamond$$

Notice that the second integration by parts in this example was very similar to the one in Example 4.28. We could have simply plugged that result in. The example was worked in full to show how two integrations by parts are needed.

Let's consider for a minute what happened in the last example. To deal with the x^2 part of the integral we integrated by parts twice – reducing the power by one in each step. That means that, for example,

$$\int x^5 \cdot e^x \cdot dx$$

would require integration by parts *five times*.

Fortunately, there is a shortcut. What happens if we integrate

$$\int f(x) \cdot e^x \cdot dx$$

Choose $U = f(x)$ and $dV = e^x \cdot dx$.

Then $dU = f'(x)$ and $V = e^x$.

Integrate by parts and we get:

$$\int f(x) \cdot e^x \cdot dx = f(x)e^x - \int f'(x)e^x \cdot dx$$

Which, used correctly, is a remarkable shortcut.

Knowledge Box 4.5

Shortcut for $\int p(x)e^x \cdot dx$

Suppose that $p(x)$ is a polynomial. Then

$$\int p(x)e^x \cdot dx = \left(p(x) - p'(x) + p''(x) - p'''(x) + \cdots\right)e^x + C$$

The formula in Knowledge Box 4.5 comes from applying the formula for $\int p(x)e^x \cdot dx$ many times. Let's use the shortcut in an example.

Example 4.32   Find

$$\int x^5 \cdot e^x \cdot dx.$$

Solution:

Applying the shortcut we get that

$$\int x^5 \cdot e^x \cdot dx = \left(x^5 - 5x^4 + 20x^3 - 60x^2 + 120x - 120\right) e^x + C$$

All we need to do is take alternating signs of x^5 and its derivatives. If the polynomial is not just a power of x this is a little more complicated.

◇

Example 4.33   Find $\int (x^2 + 2x + 2)e^x \, dx.$

Solution:

$$p(x) = x^2 + 2x + 2$$

$$p(x)' = 2x + 2$$

$$p(x)'' = 2$$

So

$$\int (x^2 + 2x + 2)e^x = (x^2 + 2x + 2 - (2x + 2) + 2)e^x + C$$

$$= (x^2 + 2)e^x + C$$

The key fact is that, if you keep taking derivatives of a polynomial, then at some point you get to zero. With a non-polynomial function the shortcut can produce an infinite object.

◇

Now we look at another technique, **circular integration by parts**. You need this technique when, in the course of integrating by parts, you arrive back at some version of the integral you started with. This sounds a bit awful, but it is actually good news. When this happens the integral may be finished using only algebra.

Example 4.34   Compute

$$\int \sin(x)e^x \cdot dx.$$

Solution:

There is no natural choice of parts. So pick anything and plunge in.

Let $U = \sin(x)$ and $dV = e^x \cdot dx$. Then $dU = \cos(x) \cdot dx$ and $V = e^x$.

$$\int \sin(x)e^x \cdot dx = \sin(x)e^x - \int \cos(x)e^x \cdot dx$$

Let: $U = \cos(x)$ and $dV = e^x \cdot dx$ (Second integration by parts.)

Then: $dU = -\sin(x) \cdot dx$ and $V = e^x$

$$= \sin(x)e^x - \left(\cos(x)e^x - \int -\sin(x)e^x \cdot dx \right)$$

$$\int \sin(x)e^x \cdot dx = (\sin(x) - \cos(x))e^x - \int \sin(x)e^x \cdot dx$$

Notice that the remaining integral equals the left-hand side.

This is the circular part.

So, $2 \int \sin(x)e^x \cdot dx = (\sin(x) - \cos(x))e^x$

Divide by two and we get:

$$\int \sin(x)e^x \cdot dx = \frac{1}{2}(\sin(x) - \cos(x))e^x + C$$

Notice that we had to separately remember to put in the $+ C$, because we finessed the final integral where we would normally have generated it.

The power of these methods of integration are expanded in the homework problems. We will also see a horrific example of circular integration by parts in Section 4.3.

PROBLEMS

Problem 4.35   Do each of the following integrals using integration by parts.

1. $\int x \cdot \sin(2x) \cdot dx$

2. $\int x \cdot e^{5x} \cdot dx$

3. $\int \ln(x) \cdot dx$

4. $\int x \cdot \ln(x) \cdot dx$

5. $\int x \cdot \sin(x) \cos(x) dx$

6. $\int \ln(x^2 + 1) \cdot dx$

Problem 4.36   For Examples 4.28-4.31, verify the formulas by taking an appropriate derivative.

Problem 4.37   Compute $\int x^n \ln(x) \cdot dx$ for $n \geq 1$

Problem 4.38   In this section we develop a shortcut for $\int p(x)e^x \cdot dx$ where $p(x)$ is a polynomial. Find the corresponding shortcut for

$$\int p(x)e^{-x} \cdot dx$$

Problem 4.39   In this section we develop a shortcut for $\int p(x)e^x \cdot dx$ where $p(x)$ is a polynomial. Find the corresponding shortcut for

$$\int p(x)\sin(x) \cdot dx$$

Problem 4.40   In this section we develop a shortcut for $\int p(x)e^x \cdot dx$ where $p(x)$ is a polynomial. Find the corresponding shortcut for

$$\int p(x)\cos(x) \cdot dx$$

Problem 4.41   Compute the following integrals using integration by parts.

1. $\int x^4 e^x \cdot dx$

2. $\int (x^3 + 2x^2 + 3x + 4)e^x \cdot dx$

3. $\int x^3 e^{-x} \cdot dx$

4. $\int (x^2 + 1) e^{-x} \cdot dx$

5. $\int x^2 \sin(x) \cdot dx$

6. $\int x^5 \cos(x) \cdot dx$

Problem 4.42   Compute the following integrals using integration by parts.

1. $\int \cos(x)e^x \cdot dx$

2. $\int \sin(2x)e^x \cdot dx$

3. $\int \cos(x)e^{3x} \cdot dx$

4. $\int \sin(x)e^{-x} \cdot dx$

5. $\int (\cos(x) + \sin(x)) e^x \cdot dx$

6. $\int \cos(2x)e^{5x} \cdot dx$

Problem 4.43   Compute

$$\int x \cdot \tan^{-1}(x) \cdot dx$$

Problem 4.44   Compute

$$\int x^3 \cdot (x^2 + 1)^{10} \cdot dx$$

Problem 4.45   Compute

$$\int x \left(\cos(ax) + \sin(bx)\right) \cdot dx$$

Problem 4.46   Compute

$$\int \sin(ax)e^{bx} \cdot dx$$

4.3   INTEGRATING TRIG FUNCTIONS

For completeness we start with a Knowledge Box of the integrals of trigonometric functions we have already obtained. Most of this section is about how to deal with products of powers of trig functions, and these integrals are building blocks.

Knowledge Box 4.6

Trig-related integral forms

- $\int \sin(x) \cdot dx = -\cos(x) + C$

- $\int \cos(x) \cdot dx = \sin(x) + C$

- $\int \tan(x) \cdot dx = \ln|\sec(x)| + C$

- $\int \sec(x) \cdot dx = \ln|\tan(x) + \sec(x)| + C$

- $\int \sec^2(x) \cdot dx = \tan(x) + C$

- $\int \csc^2(x) \cdot dx = -\cot(x) + C$

- $\int \sec(x)\tan(x) \cdot dx = \sec(x) + C$

- $\int \csc(x)\cot(x) \cdot dx = -\csc(x) + C$

Odd powers of sine and cosine

Functions of this sort are integrated by exploiting the Pythagorean identity. We leave one power of the trig function to be dU and transform the remaining even powers via one of

$$\cos^2(x) = 1 - \sin^2(x)$$

or

$$\sin^2 = 1 - \cos^2(x)$$

as appropriate, transforming the problem into an integral of a polynomial function.

Example 4.47   Find $\int \cos^5(x) \cdot dx$.

Solution:

$$\int \cos^5(x) \cdot dx = \int \left(\cos^2(x)\right)^2 \cdot \cos(x) \cdot dx$$

$$= \int \left(1 - \sin^2(x)\right)^2 \cdot \cos(x) \cdot dx$$

$$\text{Let } u = \sin(x), \, du = \cos(x) \cdot dx$$

$$= \int (1 - u^2)^2 \cdot du$$

$$= \int \left(1 - 2u^2 + u^4\right) \cdot du$$

$$= u - \frac{2}{3}u^3 + \frac{1}{5}u^5 + C$$

$$= \sin(x) - \frac{2}{3}\sin^3(x) + \frac{1}{5}\sin^5(x) + C$$

$$\diamond$$

Odd powers of sine work the same way, but with the other function in a starring role.

Even powers of sine and cosine

These are much messier and rely on the power reduction identities:

$$\sin^2(x) = \frac{1 - \cos(2x)}{2}$$

$$\cos^2(x) = \frac{1 + \cos(2x)}{2}$$

They can always be used to transform an even power into a bunch of much lower powers, both odd and even. This means that even powers of trig functions often end up as furballs, but they can be done with patience and persistence.

Example 4.48   Compute

$$\int \sin^4(x) \cdot dx$$

Solution:

$$\int \sin^4(x) \cdot dx = \int \left(\sin(x)^2\right)^2 \cdot dx$$

$$= \int \left(\frac{1 - \cos(2x)}{2}\right)^2 \cdot dx$$

$$= \frac{1}{4} \int \left(1 - 2\cos(2x) + \cos^2(2x)\right) \cdot dx$$

$$= \frac{1}{4} \int \left(1 - 2\cos(2x) + \frac{1 + \cos(4x)}{2}\right) \cdot dx$$

$$= \frac{1}{8} \int \left(2 - 4\cos(2x) + 1 + \cos(4x)\right) \cdot dx$$

$$= \frac{1}{8} \int \left(3 - 4\cos(2x) + \cos(4x)\right) \cdot dx$$

$$= \frac{1}{8} \left(3x - 4\frac{1}{2}\sin(2x) + \frac{1}{4}\sin(4x)\right) + C$$

$$= \frac{3}{8}x - \frac{1}{4}\sin(2x) + \frac{1}{32}\sin(4x) + C$$

$$\diamond$$

Even powers of secant

These are pretty easy, again by hitting them with Pythagorean identities. Save one $\sec^2(x)$ to be du and apply

$$\sec^2(x) = 1 + \tan^2(x)$$

to set up the u-substitution $u = \tan(x)$.

Example 4.49   Compute

$$\int \sec^8(x) \cdot dx$$

Solution:

$$\int \sec^8(x) \cdot dx = \int \left(\sec^2(x)\right)^3 \cdot \sec^2(x) \cdot dx$$

$$= \int \left(1 + \tan^2(x)\right)^3 \cdot \sec^2(x) \cdot dx$$

$$\text{Let } u = \tan(x),\, du = \sec^2(x) \cdot dx$$

$$= \int \left(1 + u^2\right)^3 du$$

$$= \int \left(1 + 3u^2 + 3u^4 + u^6\right) du$$

$$= u + u^3 + \frac{3}{5}u^5 + \frac{1}{7}u^7 + C$$

$$= \tan(x) + \tan^3(x) + \frac{3}{5}\tan^5(x) + \frac{1}{7}\tan^7(x) + C$$

$$\Diamond$$

Odd powers of secant

These are typically a nightmare!

Example 4.50   Compute

$$\int \sec^3(x) \cdot dx$$

Solution:

$$\int \sec^3(x) \cdot dx = \int \left(1 + \tan^2(x)\right) \sec(x) \cdot dx$$

$$= \int \sec(x) \cdot dx + \int \tan^2(x) \sec(x) \cdot dx$$

$$= \ln|\sec(x) + \tan(x)| + \int \tan(x) \cdot \sec(x) \tan(x) \cdot dx$$

$$U = \tan(x); \quad dV = \sec(x) \tan(x) \cdot dx \qquad\qquad \text{Use parts}$$

$$du = \sec^2(x) \cdot dx; \quad V = \sec(x)$$

$$= \ln|\sec(x) + \tan(x)| + \sec(x) \tan(x) - \int \sec^3(x) \cdot dx$$

$$2 \int \sec^3(x) \cdot dx = \ln|\sec(x) + \tan(x)| + \sec(x) \tan(x) \qquad\qquad \text{Circularize!}$$

$$\int \sec^3(x) \cdot dx = \frac{1}{2} \left(\ln|\sec(x) + \tan(x)| + \sec(x) \tan(x)\right) + C$$

As with all circular integrations by parts, we need to remember to finish with a $+ C$. Higher odd powers are worse.

$$\Diamond$$

Powers of tangent

Even powers of tangent can be transformed into even powers of secant by using $\tan^2(x) = \sec^2(x) - 1$.

Example 4.51   Compute

$$\int \tan^4(x) \cdot dx$$

Solution:

$$\int \tan^4(x) \cdot dx = \int \left(\sec^2(x) - 1\right)^2 \cdot dx$$

$$= \int \left(\sec^4(x) - 2\sec^2(x) + 1\right) \cdot dx$$

$$= \int \sec^4(x) \cdot dx - 2 \int \sec^2(x) \cdot dx + \int dx$$

$$= \int (\tan^2 + 1)\sec^2 \cdot dx - 2 \int \sec^2(x) \cdot dx + \int dx$$

Let: $u = \tan(x)$, $du = \sec^2(x) \cdot dx$

$$= \int (u^2 + 1) \cdot du - 2 \int \sec^2(x) \cdot dx + \int dx$$

$$= \frac{1}{3}u^3 + u - 2\tan(x) + x + C$$

$$= \frac{1}{3}\tan^3(x) + \tan(x) - 2\tan(x) + x + C$$

$$= \frac{1}{3}\tan^3(x) - \tan(x) + x + C$$

◇

Odd powers of tangent can be reduced to lower odd powers of tangent. Assume $2n + 1 > 1$. Then:

$$\int \tan^{2n+1} \cdot dx = \int (\sec^2(x) - 1) \tan^{2n-1}(x) \cdot dx$$

$$= \int \tan^{2n-1}(x) \sec^2(x) \cdot dx - \int \tan^{2n-1}(x) \cdot dx$$

Let: $u = \tan(x)$, $du = \sec^2(x) \cdot dx$

$$= \int u^{2n-1} \cdot du - \int \tan^{2n-1}(x) \cdot dx$$

$$= \frac{1}{2n} u^{2n} - \int \tan^{2n-1}(x) \cdot dx$$

$$= \frac{1}{2n} \tan^{2n}(x) - \int \tan^{2n-1}(x) \cdot dx$$

Since we also know that

$$\int \tan(x) \cdot dx = \ln|\sec(x)| + C$$

we can integrate any odd power of tangent by applying the above formula until the power is down to one.

Example 4.52   Find $\int \tan^5(x) \cdot dx$.

Solution:

$$\int \tan^5(x) \cdot dx = \frac{1}{4} \tan^4(x) - \int \tan^3(x) \cdot dx$$

$$= \frac{1}{4} \tan^4(x) - \left(\frac{1}{2} \tan^2(x) - \int \tan(x) \cdot dx \right)$$

$$= \frac{1}{4} \tan^4(x) - \frac{1}{2} \tan^2(x) + \ln|\sec(x)| + C$$

$\Diamond$

Mixed functions

When you wish to integrate a mixed product of trig functions, one strategy is to turn everything into sines and cosines and look for a good trig substitution. The fact that $u = \cos(x) \implies du = -\sin(x) \cdot dx$ and $u = \sin(x) \implies du = \cos(x) \cdot dx$ permits us to transform the trig integral into the integral of a rational function.

Example 4.53   Find

$$\int \sin(x) \tan^2(x) \cdot dx$$

Solution:

$$\int \sin(x) \tan^2(x) \cdot dx = \int \frac{\sin^3(x)}{\cos^2(x)} \cdot dx$$

$$= \int \frac{1 - \cos^2(x)}{\cos^2(x)} \cdot \sin(x) \cdot dx$$

$$\text{Let } u = \cos(x), \, du = -\sin(x) \cdot dx$$

$$= \int \frac{1 - u^2}{u^2} \cdot (-du)$$

$$= -\int \left(\frac{1}{u^2} - 1 \right) du$$

$$= -\left(-\frac{1}{u} - u \right) + C$$

$$= \frac{1}{u} + u + C = \frac{1}{\cos(x)} + \cos(x) + C$$

$$= \sec(x) + \cos(x) + C$$

$$\Diamond$$

The mainstay of the techniques for integrating trig functions presented in this section are the application of trigonometric identities. There are literally an infinite number of problems and identities that can be used to solve them. The techniques of u-substitution and integration by parts also appear, arising naturally when we try to integrate trigonometric functions.

Many of the transformations created by u-substitution make the trig function integrals into rational and even polynomial function integrals. What, however, is the point of learning to integrate these functions? In the next section we will learn to transform integrals involving square roots into trigonometric integrals. Integration is one of the best examples of how mathematics builds on itself.

PROBLEMS

Problem 4.54   Perform the following integrals.

1. $\int \cos^3(x) \cdot dx$

2. $\int \sin^5(x) \cdot dx$

3. $\int \cos^7(x) \cdot dx$

4. $\int \cos^4(x) \cdot dx$

5. $\int \sin^6(x) \cdot dx$

6. $\int \sin^2(x) \cos^2(x) \cdot dx$

Problem 4.55   Perform the following integrals.

1. $\int \sec^6(x) \cdot dx$

2. $\int \tan^6(x) \cdot dx$

3. $\int \tan^3(x) \sec^2(x) \cdot dx$

4. $\int \tan^7(x) \cdot dx$

5. $\int \cot^6(x) \cdot dx$

6. $\int \sec^3(x) \tan^3(x) \cdot dx$

Problem 4.56   Compute

$$\int \sec^5(x) \cdot dx$$

Problem 4.57   Find the power reduction integral for

$$\int \cot^{2n+1}(x) \cdot dx$$

that is analogous to the odd powers of tangent technique in this section.

Problem 4.58   Compute

$$\int \csc^3(x) \cdot dx$$

Problem 4.59   Perform the following integrals.

1. $\displaystyle\int \cos^4(x)\sin^3(x)\cdot dx$ 4. $\displaystyle\int \sec^3(x)\sin^3(x)\cdot dx$

2. $\displaystyle\int \cos^6(x)\tan^2(x)\sin(x)\cdot dx$ 5. $\displaystyle\int \frac{\tan^2(x)}{\sin(x)}\cdot dx$

3. $\displaystyle\int \cot(x)\csc^3(x)\cdot dx$ 6. $\displaystyle\int \tan^5(x)sec(x)\cdot dx$

Problem 4.60   Find $\displaystyle\int (\cos(x)+1)^5\cdot dx$

Problem 4.61   Find $\displaystyle\int \sin^2(x)\cdot\cos^5(x)\cdot dx$

Problem 4.62   Find $\displaystyle\int \sin^2(x)\cdot\cos^4(x)\cdot dx$

Problem 4.63   Find $\displaystyle\int \sin^4(x)\cdot\cos^5(x)\cdot dx$

Problem 4.64   Find $\displaystyle\int \left(\sin^2(x)+1\right)^8\cos^3(x)\cdot dx$

Problem 4.65   Find $\displaystyle\int \left(\cos^2(x)+1\right)^8\sin^3(x)\cdot dx$

Problem 4.66   Find $\displaystyle\int \sin(x)\cos(x)\cdot dx$

CHAPTER 5

Methods of Integration II

The integration methods in this chapter use more sophisticated geometry and algebra than those in the previous chapter. Both trigonometric substitution and partial fractions come up in the natural sciences and so merit inclusion in this text. They are placed in their own chapter because, with the possible exception of the material on circular integration by parts, they are more mathematically challenging.

5.1   TRIGONOMETRIC SUBSTITUTION

We have the following integrals that use inverse trig functions:

$$\int \frac{dx}{x^2 + 1} \cdot dx = \tan^{-1}(x) + C$$

$$\int \frac{dx}{\sqrt{1 - x^2}} \cdot dx = \sin^{-1}(x) + C$$

and

$$\int \frac{dx}{|x|\sqrt{x^2 - 1}} \cdot dx = \sec^{-1}(x) + C$$

The goal of this section is to enlarge our library of integrals of this type. Let's start with an example that shows how to use **trigonometric substitution**.

Example 5.1   Compute

$$\int \frac{dx}{\sqrt{x^2 + 1}}$$

Solution:

Examine the following picture of a carefully labeled right triangle.

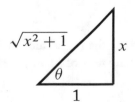

We see that $\dfrac{x}{1} = x = \tan(\theta)$, which means $dx = \sec^2(\theta) \cdot d\theta$.

Also, $\dfrac{\sqrt{x^2 + 1}}{1} = \sqrt{x^2 + 1} = \sec(\theta)$. So, we can substitute and integrate.

$$\int \frac{dx}{\sqrt{x^2 + 1}} = \int \frac{\sec^2(\theta)}{\sec(\theta)} \cdot d\theta$$

$$= \int \sec(\theta) \cdot d\theta$$

$$= \ln(|\tan(\theta) + \sec(\theta)|) + C$$

$$= \ln(|x + \sqrt{x^2 + 1}|) + C$$

For the last substitution – back to x – we need to refer back to our substitution triangle at the beginning of the integral.

When making the substitution triangle remember that it has to obey the Pythagorean theorem. If at all possible, the triangle should include all the parts in your integral.

So what just happened? Whenever you have a radical like $\sqrt{ax^2 + b}$ or $\sqrt{b - ax^2}$ in an integral, it is possible to draw a right triangle that structures a substitution for you. This substitution transforms the integral into a trigonometric integral – which goes a long way toward explaining why we worked so hard on integrating trig functions in the last chapter.

There are several types of right triangles. The next example uses a different type of right triangle from the last one. Rather than having the square root of the sum of squares

$$\sqrt{x^2 + 1}$$

on the hypotenuse of the right triangle, it will have the difference of squares

$$\sqrt{1 - x^2}$$

on one of the legs of the triangle. One of the major indicators of the type of triangle needed for substitution is the presence of the square root of a sum, as opposed to a difference, of squares.

Example 5.2   Compute

$$\int \sqrt{1-x^2} \cdot dx$$

Solution:

Here is the natural triangle for this problem:

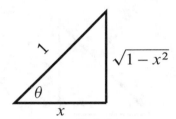

Then the substitutions are $x = \cos(\theta)$ and $\sqrt{1-x^2} = \sin(\theta)$. Taking a derivative, we see that $dx = -\sin(\theta) \cdot d\theta$. Substituting all this into the problem yields:

$$\int \sqrt{1-x^2} \cdot dx = \int \sin(\theta)(-\sin(\theta)) \cdot d\theta$$

$$= -\int \sin^2(\theta) \cdot d\theta$$

$$= -\frac{1}{2} \int (1 - \cos(2\theta)) \cdot d\theta$$

$$= \frac{1}{4} \sin(2\theta) - \frac{1}{2}\theta + C$$

But now we need to substitute back. $\sin(2\theta) = 2\sin(\theta)\cos(\theta)$ so that's not too bad. We also know that $x = \cos(\theta)$ so $\cos^{-1}(x) = \theta$. This is enough, and we get that:

$$\frac{1}{4} \sin(2\theta) - \frac{1}{2}\theta + C = \frac{1}{2} \sin(\theta)\cos(\theta) - \frac{1}{2}\theta + C$$

$$= \frac{1}{2} \left(x\sqrt{1-x^2} - \cos^{-1}(x) \right) + C$$

Done!

◊

Notice that we have a choice of which formula goes on what leg of the right triangle. Sometimes one choice leads to a slightly easier problem than the other, so it is worth thinking about which substitution you will use.

Example 5.3   Compute $\int \dfrac{x^2}{\sqrt{4-x^2}} \cdot dx$

Solution:

Here is the natural picture:

This makes the substitutions $\dfrac{x}{2} = \sin(\theta)$ and $\dfrac{\sqrt{4-x^2}}{2} = \cos(\theta)$ so that

$$\sqrt{4-x^2} = 2\cos(\theta)$$

$$x = 2\sin(\theta)$$

$$dx = 2\cos(\theta)d\theta$$

Plug all this in and integrate:

$$\int \frac{x^2}{\sqrt{4-x^2}} = \int \frac{4\sin^2(\theta)2\cos(\theta)}{2\cos(\theta)}d\theta$$

$$= 4\int \sin^2(\theta)\cdot d\theta$$

$$= 2\int (1-\cos(2\theta))\cdot d\theta$$

$$= 2\left(\theta - \frac{1}{2}\sin(2\theta)\right) + C$$

$$= 2\theta - \sin(2\theta) + C$$

As before, $\theta = \sin^{-1}\left(\dfrac{x}{2}\right)$ and $\sin(2\theta) = 2\sin(\theta)\cos(\theta) = 2x\sqrt{4-x^2}$ so

$$\int \frac{x^2}{\sqrt{4-x^2}} \cdot dx = 2\sin^{-1}\left(\frac{x}{2}\right) - 2x\sqrt{4-x^2} + C$$

$$\Diamond$$

Example 5.4   Directly verify the area formula for a circle by integration using Cartesian coordinates.

Solution:

It would be very easy to just verify the formula using polar coordinates, but for this example we are using Cartesian coordinates. Since a circle of radius R centered at the origin has the form $x^2 + y^2 = R^2$, our strategy is to take the function that graphs as the upper half of the circle

$$y = \sqrt{R^2 - x^2}$$

integrate it and double the result. Since it is an even function, we can actually do the problem

$$\text{Area} = 4\int_0^R \sqrt{R^2 - x^2} \cdot dx$$

Here is the substitution triangle:

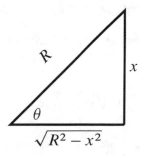

So the substitutions are:

$$\sqrt{R^2 - x^2} = R\cos(\theta)$$

$$x = R\sin(\theta)$$

$$dx = R\cos(\theta) \cdot d\theta$$

When $x = R$, we have that $\theta = \frac{\pi}{2}$. When $x = 0$, we have that $\theta = 0$. This gives us the following definite integral.

$$\text{Area} = 4 \int_0^{\pi/2} R\cos(\theta) \cdot R\cos(\theta) \cdot d\theta$$

$$= 4R^2 \int_0^{\pi/2} \cos^2(\theta) \cdot d\theta$$

$$= 2R^2 \int_0^{\pi/2} (1 + \cos(2\theta)) \cdot d\theta$$

$$= 2R^2 \left(\theta + \frac{1}{2}\sin(2\theta) \right) \Big|_0^{\pi/2}$$

$$= 2R^2 \left(\frac{\pi}{2} + \frac{1}{2}\sin(\pi) - 0 - 0 \right)$$

$$= \pi R^2$$

Which verifies the formula.

The next example will draw on a formula we already have – and paid for with blood:

$$\int \sec^3(u) \cdot du = \frac{1}{2}\left(\sec(u)\tan(u) + \ln(|\sec(u) + \tan(u)|) \right) + C$$

Example 5.5   Compute

$$\int \sqrt{x^2 + 1} \cdot dx$$

Solution:

For this one we get to re-use the first triangle picture:

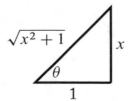

So the substitutions are:

$$\sqrt{x^2 + 1} = \sec(\theta)$$

$$x = \tan(\theta)$$

$$dx = \sec^2(\theta)d\theta$$

Substituting we get:

$$\int \sqrt{x^2 + 1} \cdot dx = \int \sec(\theta) \cdot \sec^2(\theta) \cdot d\theta$$

$$= \int \sec^3(\theta) \cdot d\theta$$

Use the known result

$$= \sec(\theta)\tan(\theta) + \ln(|\sec(\theta) + \tan(\theta)|) + C$$

$$= x\sqrt{x^2 + 1} + \ln(|x + \sqrt{x^2 + 1}|) + C$$

Without the known form, this one is a bear.

◇

Example 5.6   Compute

$$\int \frac{\sqrt{9 - x^2}}{x^2} \cdot dx$$

Solution:

Use the triangle:

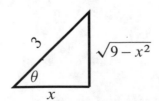

This gives us the following substitutions.

$$\sqrt{9 - x^2} = 3\sin(\theta)$$

$$x = 3\cos(\theta)$$

$$dx = -3\sin(\theta) \cdot d\theta$$

Which permits us to integrate.

$$\int \frac{\sqrt{9 - x^2}}{x^2} \cdot dx = \int \frac{3\sin(\theta) \cdot (-3\sin(\theta))}{9\cos^2(\theta)} \cdot d\theta$$

$$= -\int \frac{\sin^2(\theta)}{\cos^2(\theta)} \cdot d\theta$$

$$= -\int \tan^2(\theta) \cdot d\theta$$

$$= -\int \left(\sec^2(\theta) - 1\right) \cdot d\theta$$

$$= -\tan(\theta) + \theta + C$$

We can pull $\tan(\theta)$ directly off the triangle and, by the usual trick, $\theta = \cos^{-1}(x/3)$ making the result of substituting back to x:

$$-\frac{\sqrt{9 - x^2}}{x} + \cos^{-1}(x/3) + C$$

$\Diamond$

The next example looks at a power of a radical.

Example 5.7   Compute

$$\int \left(1 - x^2\right)^{3/2} \cdot dx$$

Solution:

We get to re-use one of our earlier triangles:

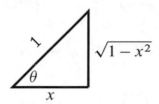

Making the substitutions:

$$\sqrt{1 - x^2} = \sin(\theta)$$

$$x = \cos(\theta)$$

$$dx = -\sin(\theta) \cdot d\theta$$

Moving on to the integral,

$$\int \left(1 - x^2\right)^{3/2} \cdot dx = \int \sin^3(\theta) \cdot (-\sin(\theta)) \cdot d\theta$$

$$= -\int \sin^4(\theta) \cdot d\theta$$

Which is Example 4.48.

$$= \frac{3}{8}\theta - \frac{1}{4}\sin(2\theta) + \frac{1}{32}\sin(4\theta) + C$$

$$= \frac{3}{8}\theta - \frac{1}{2}\sin(\theta)\cos(\theta) + \frac{1}{16}\sin(2\theta)\cos(2\theta) + C$$

$$= \frac{3}{8}\theta - \frac{1}{2}\sin(\theta)\cos(\theta) + \frac{1}{8}\sin(\theta)\cos(\theta)\left(\cos^2(\theta) - \sin^2(\theta)\right) + C$$

Using $\theta = \cos^{-1}(x)$

$$= \frac{3}{8}\cos^{-1}(x) - \frac{1}{2}x\sqrt{1-x^2} + \frac{1}{8}x\sqrt{1-x^2}(x^2 - (1-x^2)) + C$$

$$= \frac{3}{8}\cos^{-1}(x) - \frac{1}{8}\left(4x\sqrt{1-x^2} + x\sqrt{1-x^2}(2x^2 - 1)\right) + C$$

$$= \frac{3}{8}\cos^{-1}(x) - \frac{1}{8}\left(4x\sqrt{1-x^2} + \sqrt{1-x^2}(2x^3 - x)\right) + C$$

$$= \frac{3}{8}\cos^{-1}(x) - \frac{1}{8}\left(\sqrt{1-x^2}(4x + 2x^3 - x)\right) + C$$

$$= \frac{3}{8}\cos^{-1}(x) - \frac{1}{8}\sqrt{1-x^2}\left(2x^3 + 3x\right) + C$$

$$= \frac{3}{8}\cos^{-1}(x) - \frac{x}{8}\sqrt{1-x^2}\left(2x^2 + 3\right) + C$$

$$\Diamond$$

The really challenging part of this example was the back-substitution. We used two trig identities to get to where we could back-substitute:

$$\sin(2\theta) = 2\sin(\theta)\cos(\theta)$$

$$\cos(2\theta) = \cos^2(\theta) - \sin^2(\theta)$$

Trigonometric substitution is useful for square roots of sums and differences of squares. This is because these fit well onto the three sides of a right triangle. This permits us to translate to the domain of trigonometry where we can use identities to manipulate the form of the integral, often into something doable.

The possible types of triangles
Shown below are the three possible types of trianges used in trigonometric substitution. The second and third are equivalent, in a sense, but they spawn complementary trig functions when used in substution.

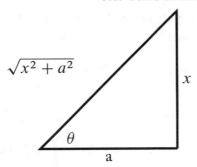

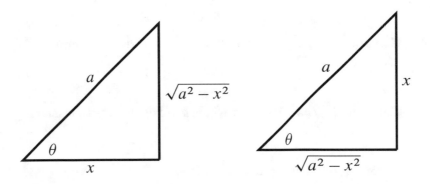

PROBLEMS

Problem 5.8   For each of the following, show a substitution triangle that might work well. Do **not** do the integrals.

1. $\displaystyle\int x^5\sqrt{1+4x^2}\cdot dx$

2. $\displaystyle\int \frac{\sqrt{4-9x^2}}{x^3}\cdot dx$

3. $\displaystyle\int (x^2+2x+1)\sqrt{x^2+2x}\cdot dx$

4. $\displaystyle\int \frac{x^2}{\sqrt{2-3x^2}}\cdot dx$

5. $\displaystyle\int \frac{x}{\sqrt{x^2+1}}\cdot dx$

6. $\displaystyle\int \frac{\sqrt{x^2+1}}{x^3}\cdot dx$

Problem 5.9   Perform the following integrals. Not all require trig substitution, but for those that do, show the substitution triangle.

1. $\displaystyle\int x\sqrt{x^2+9}\cdot dx$

2. $\displaystyle\int x^2\sqrt{x^2+1}\cdot dx$

3. $\displaystyle\int x^3\sqrt{x^2+4}\cdot dx$

4. $\displaystyle\int \frac{x^3}{\sqrt{9x^2+16}}\cdot dx$

5. $\displaystyle\int \sqrt{1+4x^2}\cdot dx$

6. $\displaystyle\int \frac{x^2}{(x^2+1)^{3/2}}\cdot dx$

Problem 5.10   Compute $\displaystyle\int \frac{dx}{\sqrt{ax^2+b^2}}$ where a,b are non-zero constants.

Problem 5.11   Compute $\displaystyle\int \sqrt{ax^2+b^2}\cdot dx$ where a,b are non-zero constants.

Problem 5.12   Perform the following integrals.

1. $\displaystyle\int \frac{\sqrt{1-x^2}}{x}\cdot dx$

2. $\displaystyle\int \frac{\sqrt{4-x^2}}{x^2}\cdot dx$

3. $\displaystyle\int \frac{dx}{\sqrt{9-4x^2}}\cdot dx$

4. $\displaystyle\int x^3\sqrt{1-x^2}\cdot dx$

5. $\displaystyle\int \frac{x^2}{\sqrt{16-x^2}}\cdot dx$

6. $\displaystyle\int \frac{x^3}{\sqrt{9-x^2}}\cdot dx$

Problem 5.13   Compute $\displaystyle\int \frac{dx}{\sqrt{a^2-b^2x^2}}\cdot dx$ where a,b are non-zero constants.

Problem 5.14   Compute $\displaystyle\int \sqrt{a^2-b^2x^2}\cdot dx$ where a,b are non-zero constants.

Problem 5.15   Perform the following integrals.

1. $\int \sqrt{\sin^2(x) + 1} \cdot \cos(x) \cdot dx$

2. $\int e^{3x} \sqrt{e^{2x} + 4} \cdot dx$

3. $\int (x + 1) \sqrt{x^2 + 2x + 2} \cdot dx$

4. $\int \sqrt{1 - e^{2x}} \cdot dx$

5. $\int \frac{\sqrt{1 - \tan^2(x)}}{\tan(x)} \sec^2(x) \cdot dx$

6. $\int x \cdot \sqrt{1 - x^2} \sqrt{1 + x^2} \cdot dx$

Problem 5.16   Compute

$$\int x\sin^{-1}(x) \cdot dx$$

5.2   PARTIAL FRACTIONS

This section introduces a novel algebraic technique for setting up integrals. The first thing we need is a useful fact about polynomials over the real numbers.

Knowledge Box 5.1

Factorization of polynomials

The only polynomials that do not factor over the real numbers are quadratics with no roots. Any polynomial of degree 3 or more factors into linear polynomials and quadratics that do not factor.

What can we do with this fact? Let's look at an example to set the stage.

Example 5.17   Compute

$$\int \frac{dx}{x^2 - 3x + 2}$$

Solution:

Notice that

$$\frac{1}{x - 2} - \frac{1}{x - 1} = \frac{(x - 1) - (x - 2)}{(x - 1)(x - 2)} = \frac{1}{x^2 - 3x + 2}$$

This means that:

$$\int \frac{dx}{x^2 - 3x + 2} = \int \frac{dx}{x - 2} - \int \frac{dx}{x - 1}$$

$$= \ln|x - 2| - \ln|x - 1| + C$$

The problem with this solution is that "notice" is not a recognized algebra technique. In order to gain the ability to "notice" in this fashion we need one additional fact. **Equal polynomials have equal coefficients**.

Suppose that

$$\frac{A}{x - 1} + \frac{B}{x - 2} = \frac{1}{(x - 1)(x - 2)}$$

Then, if we cross-multiply on the left side of the equation, we get

$$\frac{A(x - 2) + B(x - 1)}{(x - 1)(x - 2)} = \frac{1}{(x - 1)(x - 2)}$$

which simplifies to

$$(A + B)x - 2A - B = 1 = 0x + 1$$

Since these polynomials are equal we get the system of equations:

$$
\begin{aligned}
A + B &= 0 \\
-2A - B &= 1 \\
A &= -1 && \text{Second equation plus first.} \\
B &= 1 && \text{Substitute A}= -1 \text{ into the first equation.}
\end{aligned}
$$

So $A = -1$ and $B = 1$ tell us the thing we noticed:

$$\frac{1}{x - 2} - \frac{1}{x - 1} = \frac{1}{x^2 - 3x + 2}$$

$$\diamond$$

This is the basis of **integration by partial fractions**. The fact that a polynomial factors into linear and unfactorable quadratic terms means that we can always break a rational function into (i) polynomial terms and (ii) rational functions of the form

$$\frac{A}{x - u} \quad \text{or} \quad \frac{Bx + C}{x^2 + sx + t}$$

Integration by partial fractions is used to integrate rational functions.

Knowledge Box 5.2

Steps of integration by partial fractions

1. *If the numerator of the rational function is not lower degree than the denominator, divide to transform the problem into a polynomial plus a rational function with a higher degree denominator.*

2. *Factor the denominator.*

3. *Set the rational function equal to the sum of partial fraction terms.*

4. *Clear the denominator by cross multiplication.*

5. *Solve for the coefficients of the partial fraction terms based on the equal polynomials resulting from cross multiplication.*

6. *Perform the resulting integrals.*

If a factor of the denominator is repeated, it gets one partial fraction term for each power of the denominator. Let's work some examples.

Example 5.18   Find

$$\int \frac{dx}{x^2 - 4x + 3}$$

Solution:

The numerator is already lower degree. Factor the denominator. $x^2 - 4x + 3 = (x - 3)(x - 1)$ which gives us

$$\frac{1}{x^2 - 4x + 3} = \frac{A}{x - 1} + \frac{B}{x - 3}$$

meaning that

$$\frac{1}{x^2 - 4x + 3} = \frac{A(x - 3) + B(x - 1)}{x^2 - 4x + 3}$$

yielding the polynomial equality $1 = (A + B)x - (3A + B)$. We get two equations – one for the coefficient of x and one for the constant term:

$$3A + B = -1 \qquad \text{for constant term}$$
$$A + B = 0 \qquad \text{for coefficient of } x$$
$$A = -B$$
$$-2B = -1$$
$$B = 1/2$$
$$A = -1/2$$

So

$$\frac{1}{x^2 - 4x + 3} = \frac{1}{2}\left(\frac{1}{x - 3} - \frac{1}{x - 1}\right)$$

Setting up the integral:

$$\int \frac{dx}{x^2 - 4x + 3} = \frac{1}{2}\int \left(\frac{1}{x - 3} - \frac{1}{x - 1}\right) dx$$
$$= \frac{1}{2}\left(\ln|x - 3| - \ln|x - 1|\right) + C$$

$$\Diamond$$

Example 5.19 Compute

$$\int \frac{2x + 5}{x^2 - x}$$

Solution:

The numerator is already lower degree. Factor the denominator and set up the partial fractions.

$$\frac{2x + 5}{x(x - 1)} = \frac{A}{x} + \frac{B}{x - 1}$$

Cross multiply and extract the polynomials equations.

$$2x + 5 = A(x - 1) + Bx$$

which gives us the linear system

$$A + B = 2$$
$$5 = -A$$
$$A = -5$$
$$B = 7$$

So

$$\int \frac{2x+5}{x^2-x} = \int \frac{7}{x-1}dx - \int \frac{5}{x}\cdot dx$$

$$= 7\ln|x-1| - 5\ln|x| + C$$

◊

The next example will be our first with a repeated root, letting us showcase the fact that partial fractions have one part for each point of multiplicity of a root.

Example 5.20   Compute

$$\int \frac{dx}{x^3-x^2}$$

Solution:

We begin, as always, by factoring: $x^3 - x^2 = x^2(x-1)$. The root $x = 0$ has multiplicity two. This means that both x and x^2 are factors. So, the partial fraction decomposition is:

$$\frac{1}{x^3-x^2} = \frac{A}{x} + \frac{B}{x^2} + \frac{C}{x-1}$$

So, once we clear the denominator ($= x^2(x-1)$) we get

$$1 = Ax(x-1) + B(x-1) + Cx^2$$

or

$$(A+C)x^2 + (B-A)x - B = 0x^2 + 0x + 1$$

The resulting system of equations is:

$$A + C = 0$$
$$B - A = 0$$
$$-B = 1$$

So $B = -1$, $A = -1$, and $C = 1$. Now we can integrate:

$$\int \frac{dx}{x^3-x^2} = \int \left(-\frac{1}{x} - \frac{1}{x^2} + \frac{1}{x-1}\right)\cdot dx$$

$$= -\ln|x| + \frac{1}{x} + \ln|x-1| + C$$

◊

Next we look at a shortcut that permits us to, in some cases, avoid having to solve the system of equations for the coefficients of the partial fraction decomposition. Instead, we plug the roots into the equations. May terms zero out and, often, we get the values of the partial fraction coefficients directly.

Example 5.21   Compute

$$\int \frac{dx}{x^3 - 6x + 11x - 6}$$

Solution:

Start by factoring: $x^3 - 6x + 11x - 6 = (x-1)(x-2)(x-3)$ which gives us a partial fractions form:

$$\frac{1}{x^3 - 6x + 11x - 6} = \frac{A}{x-1} + \frac{B}{x-2} + \frac{C}{x-3}$$

Clearing the denominator we get

$$1 = A(x-2)(x-3) + B(x-1)(x-3) + C(x-1)(x-2)$$

which sets us up for the shortcut:

$$1 = A(-1)(-2) = 2A \qquad\qquad \text{When } x = 1$$

$$1 = B(1)(-1) = -B \qquad\qquad \text{When } x = 2$$

$$1 = C(2)(1) = 2C \qquad\qquad \text{When } x = 3$$

So $A = 1/2$, $B = -1$, and $C = 1/2$. Integrate:

$$\int \frac{dx}{x^3 - 6x + 11x - 6} = \int \left(\frac{1/2}{x-1} - \frac{1}{x-2} + \frac{1/2}{x-3} \right) \cdot dx$$

$$= \frac{1}{2} \ln|x-1| - \ln|x-2| + \frac{1}{2} \ln|x-3| + C$$

◊

What happened? Let's summarize our findings in a Knowledge Box.

Knowledge Box 5.3

Finding the coefficients of a partial fractions decomposition

- *Equal polynomials have equal coefficients for equal power terms.*

- *Equal polynomials are equal for any specific value of their variable.*

The shortcut for finding the partial fraction constants is to plug in values of x that zero out all but one term. Warning: this isn't always possible. Terms with multiplicity above one or non-factorable quadratic terms can mess up the shortcut. The next set of examples involve integrating reciprocals of quadratics that don't factor. These will arise as part of partial fraction decompositions.

Example 5.22   Compute

$$\int \frac{dx}{x^2 + 4x + 5}$$

Solution:

The denominator doesn't factor. Let's try completing the square.

$$\int \frac{dx}{x^2 + 4x + 5} = \int \frac{dx}{x^2 + 4x + 4 + 1}$$

$$= \int \frac{dx}{(x + 2)^2 + 1}$$

Let $u = x + 2$, so $du = dx$

$$= \int \frac{du}{u^2 + 1}$$

$$= \tan^{-1}(u) + C$$

$$= \tan^{-1}(x + 2) + C$$

◇

Knowledge Box 5.4

Integrating the reciprocal of unfactorable quadratics

1. *Complete the square to place the quadratic in the form* $a(x - b)^2 + c$

2. $\dfrac{1}{a(x - b)^2 + c} = \dfrac{1}{c\left(\frac{a}{c}(x - b)^2 + 1\right)} = \dfrac{1}{c\left(\left(\sqrt{\frac{a}{c}}(x - b)\right)^2 + 1\right)}$

3. *Let* $u = \sqrt{\dfrac{a}{c}}(x - b)$ *so that* $du = \sqrt{\dfrac{a}{c}}dx$ *and* $dx = \sqrt{\dfrac{c}{a}}du$

4. $\displaystyle\int \frac{1}{a(x - b)^2 + c} \cdot dx = \frac{1}{c}\sqrt{\frac{c}{a}}\tan^{-1}(u) + C$

$$= \frac{1}{\sqrt{ac}}\tan^{-1}\left(\sqrt{\frac{a}{c}}(x - b)\right) + C$$

With this particular type of u-substitution in place we can do the next batch of integrations by partial fractions.

Example 5.23   Compute

$$\int \frac{dx}{x^3 + x}$$

Solution:

Factor and we see that $x^3 + x = x(x^2 + 1)$. So the partial fraction form is:

$$\frac{1}{x^3 + x} = \frac{A}{x} + \frac{Bx + C}{x^2 + 1}$$

We need $Bx + C$ in the numerator of the second part instead of just B because $x^2 + 1$ is quadratic. Clearing the denominator we get the polynomial equation:

$$1 = A(x^2 + 1) + x(Bx + C)$$

Plug in $x = 0$ and we see that $1 = A$. This tells us that

$$1 = (x^2 + 1) + Bx^2 + Cx$$

or

$$-x^2 + 0x = Bx^2 + Cx$$

so $B = -1$ and and $C = 0$. We are now prepared to integrate

$$\int \frac{dx}{x^3 + x} = \int \left(\frac{1}{x} - \frac{x}{x^2 + 1} \right) \cdot dx$$

$$= \ln |x| - \frac{1}{2} \ln |x^2 + 1| + C$$

The second part of the integral is a u-substitution with $u = x^2 + 1$; we have done it before.

Let's try a less neat integral of this sort.

Example 5.24 Compute

$$\int \frac{3x \cdot dx}{x^3 - 9x^2 + 25x - 25}$$

Solution:

Factor, and we get $x^3 - 9x^2 + 25x - 25 = (x - 5)(x^2 - 4x + 5) = (x - 5)((x - 2)^2 + 1)$. That makes the partial fraction form

$$\frac{3x \cdot dx}{x^3 - 9x^2 + 25x - 25} = \frac{A}{x - 5} + \frac{Bx + C}{x^2 - 4x + 5}$$

Clear the denominator and we get

$$0x^2 + 3x + 0 = A(x^2 - 4x + 5) + (Bx + C)(x - 5) = Ax^2 - 4Ax + 5A + Bx^2 - 5Bx + Cx + 5C$$

yielding the simultaneous equations

$$A + B = 0$$
$$-4A - 5B + C = 3$$
$$5A - 5C = 0$$

So, $A = -B$, $A = C$, and $-4A + 5A + A = 2A = 3$. So, $A = C = 3/2$ and $B = -3/2$.

This permits us to integrate:

$$\int \frac{3x \cdot dx}{x^3 - 9x^2 + 25x - 25} = \frac{3}{2} \int \left(\frac{1}{x-5} - \frac{x-1}{(x-2)^2 + 1} \right) \cdot dx$$

$$= \frac{3}{2} \int \left(\frac{1}{x-5} - \frac{x-2}{(x-2)^2 + 1} + \frac{1}{(x-2)^2 + 1} \right) \cdot dx$$

Let $u = x - 2$ with $du = dx$ for the second and third terms

$$= \frac{3}{2} \left(\ln|x-5| - \frac{1}{2} \ln(u^2 + 1) + \tan^{-1}(u) \right) + C$$

$$= \frac{3}{2} \left(\ln|x-5| - \frac{1}{2} \ln(x^2 - 4x + 5) + \tan^{-1}(x-2) \right) + C$$

$$\Diamond$$

Example 5.25   Compute

$$\int \frac{dx}{x^3 - 1}$$

Solution:

Factor: $x^3 - 1 = (x-1)(x^2 + x + 1)$. Then

$$\frac{1}{x^3 - 1} = \frac{A}{x-1} + \frac{Bx + C}{x^2 + x + 1}$$

Clear the denominator and we get

$$1 = A(x^2 + x + 1) + (Bx + C)(x - 1)$$

yielding:

$$A + B = 0$$

$$A - B + C = 0$$

$$A - C = 1$$

Solving the linear system we get $A = 1/3$, $B = -1/3$, $C = -2/3$.

$$\int \frac{dx}{x^3 - 1} = \frac{1}{3} \int \left(\frac{1}{x - 1} - \frac{x + 2}{x^2 + x + 1} \right) \cdot dx$$

$$= \frac{1}{3} \int \left(\frac{1}{x - 1} - \frac{1}{2} \cdot \frac{2x + 4}{x^2 + x + 1} \right) \cdot dx$$

$$= \frac{1}{3} \int \left(\frac{1}{x - 1} - \frac{1}{2} \cdot \frac{2x + 1 + 3}{x^2 + x + 1} \right) \cdot dx$$

$$= \frac{1}{3} \int \left(\frac{1}{x - 1} - \frac{1}{2} \cdot \frac{2x + 1}{x^2 + x + 1} - \frac{1}{2} \cdot \frac{3}{x^2 + x + 1} \right) \cdot dx$$

The middle term is a u-substitution that leads to a log.

$$= \frac{1}{3} \ln|x - 1| - \frac{1}{6} \ln|x^2 + x + 1| - \frac{1}{2} \int \frac{dx}{x^2 + x + 1}$$

$$= \frac{1}{3} \ln|x - 1| - \frac{1}{6} \ln|x^2 + x + 1| - \frac{1}{2} \int \frac{dx}{(x + 1/2)^2 + 3/4}$$

$$= \frac{1}{3} \ln|x - 1| - \frac{1}{6} \ln|x^2 + x + 1| - \frac{1}{2} \int \frac{dx}{\frac{3}{4}\left(\frac{4}{3}(x + 1/2)^2 + 1\right)}$$

$$= \frac{1}{3} \ln|x - 1| - \frac{1}{6} \ln|x^2 + x + 1| - \frac{1}{2} \cdot \frac{4}{3} \int \frac{dx}{\left(\frac{4}{3}(x + 1/2)^2 + 1\right)}$$

$$= \frac{1}{3} \ln|x - 1| - \frac{1}{6} \ln|x^2 + x + 1| - \frac{2}{3} \int \frac{dx}{\left(\frac{2}{\sqrt{3}}(x + 1/2)\right)^2 + 1}$$

Let $u = \frac{2}{\sqrt{3}}(x + \frac{1}{2})$ and so $du = \frac{2}{\sqrt{3}}dx$

$$= \frac{1}{3} \ln|x - 1| - \frac{1}{6} \ln|x^2 + x + 1| - \frac{2}{3} \cdot \frac{\sqrt{3}}{2} \int \frac{du}{u^2 + 1}$$

$$= \frac{1}{3} \ln|x - 1| - \frac{1}{6} \ln|x^2 + x + 1| - \frac{1}{\sqrt{3}} \tan^{-1}(u) + C$$

$$= \frac{1}{3} \ln |x - 1| - \frac{1}{6} \ln |x^2 + x + 1| - \frac{1}{\sqrt{3}} \tan^{-1} \left(\frac{2}{\sqrt{3}} \left(x + \frac{1}{2} \right) \right) + C$$

See! Easy!

◇

One of the fairly obvious conclusions that follows from this section is that you need to not only know algebra, you need to be a master of it to mess with partial fractions in all but its simplest form. Some of the homework problems develop general formulas for this sort of super messy inverse tangent integral. Like Knowledge Box 5.4, but with more parameters.

PROBLEMS

Problem 5.26   Perform the following integrals.

1. $\displaystyle\int \frac{dx}{x^2 - x}$

2. $\displaystyle\int \frac{dx}{x^2 - 6x + 8}$

3. $\displaystyle\int \frac{dx}{x^2 + 4x + 3}$

4. $\displaystyle\int \frac{dx}{6x^2 - 5x + 1}$

5. $\displaystyle\int \frac{dx}{x^2 - x - 1}$

6. $\displaystyle\int \frac{dx}{x^2 - 11x + 30}$

Problem 5.27   Perform each of the following integrals. Partial fractions are not needed.

1. $\displaystyle\int \frac{1}{x^2 + 2x + 2} \cdot dx$

2. $\displaystyle\int \frac{1}{x^2 + 3x + 4} \cdot dx$

3. $\displaystyle\int \frac{1}{x^2 + x + 2} \cdot dx$

4. $\displaystyle\int \frac{x + 1}{x^2 + 2x + 2} \cdot dx$

5. $\displaystyle\int \frac{2x + 2}{x^2 + 3x + 4} \cdot dx$

6. $\displaystyle\int \frac{x - 1}{x^2 + x + 2} \cdot dx$

Problem 5.28   Perform each of the following integrals.

1. $\displaystyle\int \frac{dx}{x^3 + 6x^2 + 11x + 6}$

2. $\displaystyle\int \frac{dx}{x^4 - 16}$

3. $\displaystyle\int \frac{dx}{x^3 - 5x^2 - 2x + 24}$

4. $\displaystyle\int \frac{x^2 + x + 1}{x^3 - 6x^2 + 11x + 6} \cdot dx$

5. $\displaystyle\int \frac{x^2 + 1}{x^3 + 4x} \cdot dx$

6. $\displaystyle\int \frac{3x - 2}{x^3 + 2x^2 - 5x - 6} \cdot dx$

Problem 5.29   For each of the following, compute the partial fractions form for decomposing the integral, but do not solve for the partial fractions coefficients or perform the integral.

1. $\int \dfrac{dx}{x^5 + 4x^3 + x}$

2. $\int \dfrac{dx}{x^4 + 4x^3 + 7x^2 + 6x + 3}$

3. $\int \dfrac{dx}{x^5 + x^3}$

4. $\int \dfrac{dx}{x^6 - 5x^5 + 6x^4 - x^3 + 5x^2 - 5x + 1}$

5. $\int \dfrac{dx}{x^6 - 3x^5 + 3x^4 + x^3}$

6. $\int \dfrac{dx}{x^6 + 2x^5 + 5x^3 + 6x^2 + 3x + 2}$

Problem 5.30   Compute

$$\int \frac{dx}{ax^2 + bx + c}$$

There are three cases each of which requires its own method of integration.

Problem 5.31   Compute $\int \dfrac{dx}{x^3 - a^3}$ where a is a positive constant.

Problem 5.32   Compute $\int \dfrac{dx}{x^4 - a^4}$ where a is a positive constant.

Problem 5.33   Perform the integrals in Problem 5.29.

Problem 5.34   Compute

$$\int \frac{x^5}{x^2 + x + 1} \cdot dx$$

Problem 5.35   Compute

$$\int \frac{dx}{(x - a)(x - b)(x - c)} \cdot dx$$

Problem 5.36   Compute

$$\int \frac{x \cdot dx}{(x-a)(x-b)(x-c)} \cdot dx$$

Problem 5.37   Compute

$$\int \frac{x^2 \cdot dx}{(x-a)(x-b)(x-c)} \cdot dx$$

5.3   PRACTICING INTEGRATION

One of the important skills for performing integrals is figuring out which method of integration to use. Knowledge Box 5.5 may be some help (it continues on the next page). This section is no more than a collection of problems – but, unlike all the other sections, you cannot guess which method is useful by looking at the section the problem appears in. Practice! Practice! Practice!

Knowledge Box 5.5

Which method of integration do I use?

1. *Polynomials are integrated with the power rule, one term at a time.*

2. *If the thing you are integrating is the derivative of something you recognize, use the fundamental theorem – the integral is the thing you recognize, plus "C".*

3. *Look at the known derivatives, e.g., $Dx \ \arctan(x) = \frac{dx}{x^2+1}$. If you find one, apply rule "b".*

4. *Will an algebraic re-arrangement of the terms of the integral set up an integral you can do?*

5. *Look through the examples in this chapter and in Chapter 1. Are any of them similar enough to your problem to help?*

(continued)

6. *Check if there is a u–substitution that makes a simpler integral. Remember that du needs to be in there in an appropriate form.*

7. *Are there natural parts for integration by parts? Remember you may need to integrate by parts several times.*

8. *Is your integral packed with trig functions? Flip through Section 4.3 which has special methods for many different combinations of trig functions. Also remember that rule "c" includes applying trig identities to set up one of these special trig methods.*

9. *Does the integral contain $\sqrt{x^2 \pm a^2}$ or $\sqrt{a^2 \pm x^2}$? Consider trigonometric substitution. This may yield integrals covered by any of rules "b", "c", or "d".*

10. *Is the integral a ratio of polynomials? If the numerator is not lower degree, divide to get a remainder that does have a lower degree top. Integrate the resulting ratio of polynomials with partial fractions.*

11. *If one of these rules seems to make some progress, keep going, you may need many steps to finish an integral.*

12. *Try rule "d" again. Really.*

13. *There are a lot of integrals that you can't do, there are many that no one can do. If you are really stuck, check with someone who has more experience than you. They may be able to recognize impossible integrals or integrals well above your level.*

Remember that a problem may use *several* methods of integration to complete one intergral: a *u*-substitution sets up a partial fractions decomposition that leads to a trig-substitution integral.

PROBLEMS

Problem 5.38   Pick and state a method of integration and then perform the integration for each of the following problems.

1. $\displaystyle\int x^3 e^x \cdot dx$

2. $\displaystyle\int x^2 \cdot (x^3 + 8) \cdot dx$

3. $\displaystyle\int \frac{x}{x^2 + 1} \cdot dx$

4. $\displaystyle\int \frac{x}{x^3 - 1} \cdot dx$

5. $\displaystyle\int \sin(x)\cos(x) \cdot dx$

6. $\displaystyle\int \sin^4(3x) \cdot dx$

Problem 5.39   Pick and state a method of integration and then perform the integration for each of the following problems.

1. $\displaystyle\int \sin(2x)e^{3x} \cdot dx$

2. $\displaystyle\int \sqrt{25x^2 + 1} \cdot dx$

3. $\displaystyle\int \cos(x)\sin(2x) \cdot dx$

4. $\displaystyle\int \frac{\ln^3(x) + 1}{x} \cdot dx$

5. $\displaystyle\int \csc^4(x) \cdot dx$

6. $\displaystyle\int \frac{x^2 - 1}{x^2 + 1} \cdot dx$

Problem 5.40   Pick and state a method of integration and then perform the integration for each of the following problems.

1. $\displaystyle\int \frac{dx}{\sqrt{1 - 4x^2}}$

2. $\displaystyle\int \frac{x}{\sqrt{1 - 9x^2}} \cdot dx$

3. $\displaystyle\int x \cdot e^{\sqrt{x}} \cdot dx$

4. $\displaystyle\int \frac{e^x}{1 + e^{2x}} \cdot dx$

5. $\displaystyle\int \frac{e^x}{e^{2x} - 3e^x + 2} \cdot dx$

6. $\displaystyle\int x \cdot \ln(x) \cdot dx$

Problem 5.41   Pick and state a method of integration and then perform the integration for each of the following problems.

1. $\displaystyle\int \frac{x^2}{x^4 - 1} \cdot dx$

2. $\displaystyle\int x \cdot e^{\sqrt[3]{x}} \cdot dx$

3. $\displaystyle\int \frac{\cos(x)}{\sin^2(x) - 1} \cdot dx$

4. $\displaystyle\int \frac{\sin(2x)}{\cos^2(x) - 9} \cdot dx$

5. $\displaystyle\int x^5 (x^2 + 2)^4 \cdot dx$

6. $\displaystyle\int \left(\cos^2(x) - \sin^2(x)\right) e^{\sin(2x)} \cdot dx$

Problem 5.42   Compute

$$\int (\sin(x) + \cos(x))^2 \cdot dx$$

Problem 5.43   Compute

$$\int x^n \cdot \ln(x) \cdot dx$$

Problem 5.44   Compute

$$\int x e^{\sqrt[n]{x}} \cdot dx$$

Problem 5.45   Compute

$$\int (\sin(ax) + cos(bx)) \, e^x \cdot dx$$

Problem 5.46   Compute

$$\int (1 + \ln(x)) x^x \cdot dx$$

Problem 5.47   Find a problem in this section that can be done in two ways and demonstrate them.

APPENDIX A

Useful Formulas

A.1 POWERS, LOGS, AND EXPONENTIALS

RULES FOR POWERS

- $a^{-n} = \dfrac{1}{a^n}$

- $\dfrac{a^n}{a^m} = a^{n-m}$

- $a^n \times a^m = a^{n+m}$

- $(a^n)^m = a^{n \times m}$

LOG AND EXPONENTIAL ALGEBRA

- $b^{\log_b(c)} = c$

- $\log_b(x^y) = y \cdot \log_b(x)$

- $\log_b(b^a) = a$

- $\log_b(xy) = \log_b(x) + \log_b(y)$

- $\log_c(x) = \dfrac{\log_b(x)}{\log_b(c)}$

- $\log_b\left(\dfrac{x}{y}\right) = \log_b(x) - \log_b(y)$

- If $\log_b(c) = a$, then $c = b^a$

A.2 TRIGONOMETRIC IDENTITIES

TRIG FUNCTION DEFINITIONS FROM SINE AND COSINE

- $\tan(\theta) = \dfrac{\sin(\theta)}{\cos(\theta)}$

- $\tan(\theta) = \dfrac{1}{\cot(\theta)}$

- $\csc(\theta) = \dfrac{1}{\sin(\theta)}$

- $\cot(\theta) = \dfrac{\cos(\theta)}{\sin(\theta)}$

- $\sec(\theta) = \dfrac{1}{\cos(\theta)}$

PERIODICITY IDENTITIES

- $\sin(x + 2\pi) = \sin(x)$

- $\sec(x) = \csc\left(x + \frac{\pi}{2}\right)$

- $\sin(x + \pi) = -\sin(x)$

- $\cos(x + 2\pi) = \cos(x)$

- $\cos(-x) = \cos(x)$

- $\cos(x + \pi) = -\cos(x)$

- $\sin(x) = \cos\left(x - \frac{\pi}{2}\right)$

- $\sin(-x) = -\sin(x)$

- $\tan(x + \pi) = \tan(x)$

- $\tan(x) = -\cot\left(x - \frac{\pi}{2}\right)$

- $\tan(x) = -\tan(x)$

THE PYTHAGOREAN IDENTITIES

- $\sin^2(\theta) + \cos^2(\theta) = 1$ • $\tan^2(\theta) + 1 = \sec^2(\theta)$ • $1 + \cot^2(\theta) = \csc^2(\theta)$

THE LAW OF SINES, THE LAW OF COSINES

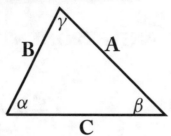

The Law of Sines

$$\frac{A}{\sin(\alpha)} = \frac{B}{\sin(\beta)} = \frac{C}{\sin(\gamma)}$$

The Law of Cosines

$$C^2 = A^2 + B^2 + 2AB \cdot \cos(\gamma)$$

The laws refer to the diagram.

SUM, DIFFERENCE, AND DOUBLE ANGLE

- $\sin(\alpha + \beta) = \sin(\alpha)\cos(\beta) + \sin(\beta)\cos(\alpha)$
- $\cos(\alpha + \beta) = \cos(\alpha)\cos(\beta) - \sin(\alpha)\sin(\beta)$
- $\sin(\alpha - \beta) = \sin(\alpha)\cos(\beta) - \sin(\beta)\cos(\alpha)$
- $\cos(\alpha - \beta) = \sin(\alpha)\sin(\beta) + \cos(\alpha)\cos(\beta)$
- $\sin(2\theta) = 2\sin(\theta)\cos(\theta)$

- $\cos(2\theta) = \cos^2(\theta) - \sin^2(\theta)$
- $\cos^2(\theta/2) = \dfrac{1 + \cos(\theta)}{2}$
- $\sin^2(\theta/2) = \dfrac{1 - \cos(\theta)}{2}$

A.3   SPEED OF FUNCTION GROWTH

- Logarithms grow faster than constants.
- Positive powers of x grow faster than logarithms.
- Larger positive powers of x grow faster than smaller positive powers of x.
- Exponentials (with positive exponents) grow faster than positive powers of x.
- Exponentials with larger exponents grow faster than those with smaller exponents.

A.4 DERIVATIVE RULES

- If $f(x) = x^n$ then
$$f'(x) = nx^{n-1}$$

- $(f(x) + g(x))' = f'(x) + g'(x)$

- $(a \cdot f(x))' = a \cdot f'(x)$

- If $f(x) = \ln(x)$, then $f'(x) = \dfrac{1}{x}$

- If $f(x) = \log_b(x)$, then $f'(x) = \dfrac{1}{x \ln(b)}$

- If $f(x) = e^x$, then $f'(x) = e^x$

- If $f(x) = a^x$, then $f'(x) = \ln(a) \cdot a^x$

- $(\sin(x))' = \cos(x)$

- $(\cos(x))' = -\sin(x)$

- $(\tan(x))' = \sec^2(x)$

- $(\cot(x))' = -\csc^2(x)$

- $(\sec(x))' = \sec(x)\tan(x)$

- $(\csc(x))' = -\csc(x)\cot(x)$

- $\left(\sin^{-1}(x)\right)' = \dfrac{1}{\sqrt{1-x^2}}$

- $\left(\cos^{-1}(x)\right)' = \dfrac{-1}{\sqrt{1-x^2}}$

- $\left(\tan^{-1}(x)\right)' = \dfrac{1}{1+x^2}$

- $\left(\cot^{-1}(x)\right)' = \dfrac{-1}{1+x^2}$

- $\left(\sec^{-1}(x)\right)' = \dfrac{1}{|x|\sqrt{x^2-1}}$

- $\left(\csc^{-1}(x)\right)' = \dfrac{-1}{|x|\sqrt{x^2-1}}$

The product rule

$$(f(x) \cdot g(x))' = f(x)g'(x) + f'(x)g(x)$$

The quotient rule

$$\left(\frac{f(x)}{g(x)}\right)' = \frac{g(x)f'(x) - f(x)g'(x)}{g^2(x)}$$

The reciprocal rule

$$\left(\frac{1}{f(x)}\right)' = \frac{-f'(x)}{f^2(x)}$$

The chain rule

$$(f(g(x)))' = f'(g(x)) \cdot g'(x)$$

ALGEBRA OF SUMMATION

- $\displaystyle\sum_{i=a}^{b} f(i) + g(i) = \sum_{i=a}^{b} f(i) + \sum_{i=a}^{b} g(i)$

- $\displaystyle\sum_{i=a}^{b} c \cdot f(i) = c \cdot \sum_{i=a}^{b} f(i)$

CLOSED SUMMATION FORMULAS

- $\displaystyle\sum_{i=1}^{n} 1 = n$

- $\displaystyle\sum_{i=1}^{n} i^2 = \frac{n(n+1)(2n+1)}{6}$

- $\displaystyle\sum_{i=1}^{n} i = \frac{n(n+1)}{2}$

- $\displaystyle\sum_{i=1}^{n} i^3 = \frac{n^2(n+1)^2}{4}$

A.5   VECTOR ARITHMETIC

VECTOR ARITHMETIC AND ALGEBRA

- $c \cdot \vec{v} = (cv_1, cv_2, \ldots, cv_n)$

- $\vec{v} + \vec{w} = (v_1 + w_1, v_2 + w_2, \ldots, v_n + w_n)$

- $\vec{v} - \vec{w} = (v_1 - w_1, v_2 - w_2, \ldots, v_n - w_n)$

- $\vec{v} \cdot \vec{w} = v_1 w_1 + v_2 w_2 + \ldots + v_n w_n$

- $c \cdot (\vec{v} + \vec{w}) = c \cdot \vec{v} + c \cdot \vec{w}$

- $c \cdot (d \cdot \vec{v}) = (cd) \cdot \vec{v}$

- $\vec{v} + \vec{w} = \vec{w} + \vec{v}$

- $\vec{u} \cdot (\vec{v} + \vec{w}) = \vec{u} \cdot \vec{v} + \vec{u} \cdot \vec{w}$

CROSS PRODUCT OF VECTORS

- $\vec{v} \times \vec{w} = (v_2 w_3 - v_3 v_2,\ v_3 w_1 - v_1 w_3,\ v_1 w_2 - v_2 w_1)$

Formula for the angle between vectors

$$\cos(\theta) = \frac{\vec{v} \cdot \vec{w}}{|v||w|}$$

A.6   POLAR AND RECTANGULAR CONVERSION

- $x = r \cdot \cos(\theta)$

- $r = \sqrt{x^2 + y^2}$

- $y = r \cdot \sin(\theta)$

- $\theta = \tan^{-1}(y/x)$

A.7   INTEGRAL RULES

BASIC INTEGRATION RULES

- $\int x^n \cdot dx = \dfrac{1}{n+1} x^{n+1} + C$

- $\int a \cdot f(x) \cdot dx = a \cdot \int f(x) \cdot dx$

- $\int (f(x) + g(x)) \cdot dx = \int f(x) \cdot dx + \int g(x) \cdot dx$

LOG AND EXPONENT

- $\int \dfrac{1}{x} \cdot dx = \ln(x) + C$

- $\int e^x \cdot dx = e^x + C$

TRIG AND INVERSE TRIG

- $\int \sin(x) \cdot dx = -\cos(x) + C$

- $\int \cos(x) \cdot dx = \sin(x) + C$

- $\int \sec^2(x) \cdot dx = \tan(x) + C$

- $\int \csc^2(x) \cdot dx = -\cot(x) + C$

- $\int \sec(x)\tan(x) \cdot dx = \sec(x) + C$

- $\int \csc(x)\cot(x) \cdot dx = -\csc(x) + C$

- $\int \dfrac{1}{\sqrt{1-x^2}} \cdot dx = \sin^{-1}(x) + C$

- $\int \dfrac{1}{1+x^2} \cdot dx = \tan^{-1}(x) + C$

- $\int \dfrac{1}{x\sqrt{x^2-1}} \cdot dx = \sec^{-1}(|x|) + C$

- $\int \tan(x) \cdot dx = \ln|\sec(x)| + C$

- $\int \sec(x) \cdot dx = \ln|\sec(x) + \tan(x)| + C$

INTEGRATION BY PARTS

$$\int U \cdot dV = UV - \int V \cdot dU$$

EXPONENTIAL/POLYNOMIAL SHORTCUT

$$\int p(x)e^x \cdot dx = \left(p(x) - p'(x) + p''(x) - p'''(x) + \cdots \right) e^x + C$$

Author's Biography

DANIEL ASHLOCK

Daniel Ashlock is a Professor of Mathematics at the University of Guelph. He has a Ph.D. in Mathematics from the California Institute of Technology, 1990, and holds degrees in Computer Science and Mathematics from the University of Kansas, 1984. Dr. Ashlock has taught mathematics at levels from 7th grade through graduate school for four decades, starting at the age of 17. Over this time Dr. Ashlock has developed a number of ideas about how to help students overcome both fear and deficient preparation. This text, covering the mathematics portion of an integrated mathematics and physics course, has proven to be one of the more effective methods of helping students learn mathematics with physics serving as an ongoing anchor and example.

Index

United States
or Publisher Services